Biomagnetic Stimulation

Biomagnetic Stimulation

Edited by

Shoogo Ueno
Kyushu University
Fukuoka, Japan

Plenum Press • New York and London

Library of Congress Cataloging-in-Publication Data

Biomagnetic stimulation / edited by Shoogo Ueno.
p. cm.
"Proceedings of an International Symposium on Biomagnetic Stimulation, held July 15, 1991, in Fukuoka, Japan"--T.p. verso.
Includes bibliographical references and index.
ISBN 0-306-44707-X
1. Brain stimulation--Congresses. 2. Brain--Magnetic fields--Congresses. 3. Heart--Magnetic fields--Congresses. I. Ueno, Shoogo. II. International Symposium on Biomagnetic Stimulation (1991 : Fukuoka-shi, Japan)
QP388.B56 1994
612'.01442--dc20 94-8276
CIP

Proceedings of an International Symposium on Biomagnetic Stimulation, held July 15, 1991, in Fukuoka, Japan

ISBN 0-306-44707-X

A Division of Plenum Publishing Corporation
233 Spring Street, New York, N.Y. 10013

Printed in the United States of America

SPONSORS

Kyushu University
World Congress on Medical Physics and Biomedical Engineering
IEEE Magnetics Society, Tokyo Chapter
IEEE Engineering in Medicine and Biology Society, Tokyo Chapter
The Japan Society of Medical Electronics and Biological Engineering
Japan Biomagnetism and Bioelectromagnetics Society
Japan Society of Applied Electromagnetics
The Institute of Electrical Engineers of Japan, Magnetics Committee
Life Support Technology Society
The Japanese Conference of Topographic EEG
The Institute of Electrical Engineers of Japan, Kyushu Chapter
The Institute of Electronics, Information and Communication Engineers, Kyushu Chapter
The Society of Instrument and Control Engineers, Kyushu Chapter
The Information Processing Society of Japan, Kyushu Chapter
The Japan Society of Medical Electronics and Biological Engineering, Kyushu Chapter

FINANCIAL CONTRIBUTORS

Celcom, Inc.
Hitachi Ltd.
IBM Japan
Japan Tobacco, Inc.
Kanematsu Electronics
Kobe Steel
Kyushu Electric Power Co.
Matsushita Electric Industrial Co.
Mitsubishi Electric Corp.
Mitsubishi Heavy Industries
NEC Corp.
NEC San-ei Instruments
NEC Software Kyushu
Nihon Kohden Corp.
Nihon Schering
Nikkiso Co.
Nippon Steel Corp.
Nippon Telegraph and Telephone Corp.
Pioneer Electronic Corp.
Saibu Gas Co.
Seiko Instruments Inc.
Siemens-Asahi Medical Systems
Simadzu Corp.
Sumitomo Metal Mining Co.
Takenaka Corp.
Teijin Limited
Toshiba Corp.
Toshiba Medical Engineering Co.
Toto Ltd.
Yokogawa Medical Systems

PREFACE

The International Symposium on Biomagnetic Stimulation was held on July 15, 1991 at the International Hall of the Hakozaki campus of Kyushu University in Fukuoka, Japan. It was a satellite symposium to the World Congress on Medical Physics and Biomedical Engineering in Kyoto, which was held July 7-11, 1991.

Successful magnetic stimulation of the human brain was first reported by Dr. Anthony Barker and his group at the University of Sheffield in the United Kingdom, in May, 1985. Of course, magnetic nerve stimulation had been studied and reported before then, but Dr. Barker's reports of successful stimulation of the brain made a strong impact on the scientific community. Since then, magnetic nerve stimulation has been widely and rapidly investigated by many groups throughout the world. This symposium focused mostly on magnetic brain stimulation.

Magnetic resonance imaging has become an indispensable technique for clinical diagnosis and medical science. The most advanced MRI techniques, such as echo planar imaging, have the potential hazard of stimulating nerve tissues due to the rapid change of gradient magnetic fields. Potential risks of MRI, including problems with gradient magnetic fields, were discussed at the symposium. Magnetic stimulation of the heart was also discussed.

In the magnetic stimulation of the brain and the heart, strong magnetic fields are usually used. Recently however, quite a few people have been devoting themselves to understanding the interaction of very weak and extremely low frequency magnetic fields with biological systems, and this topic was also taken up at the symposium.

The original plan for this book was to record the oral presentations from video tapes, to convey the conference's atmosphere to readers. But in the editing process, most speakers -- including the editor -- substantially rewrote their talks in a more formal style. The papers of two of the speakers, however, appear here in the original oral style, which the reader will no doubt appreciate. At the end of each chapter, transcriptions of the question-answer sessions are included.

This book serves as an introduction to newcomers to the field, whether they are physicians, engineers, or other scientists. At the same time, it should also provide some new information for specialists in biomagnetics.

Great appreciation is due to those who helped compile and prepare this book, especially Ellen Barton, who transcribed the proceedings from the video tape, and Susanna Heckman, who did much of the copyediting.

Heartfelt thanks, also, to the many colleagues and staff in Kyushu University's Electronics Department, who helped with either the book or the symposium or both, particularly Keiji Iramina, Tsuruo Matsuda, Osamu Hiwaki, Seiko Hamano, Fusako Tani, Kyoko Hieda and Terumi Asai.

As the symposium organizer and the editor of this book, I wish to thank the many sponsors who provided financial support for the symposium and publishing.

Shoogo Ueno
Editor

PRESIDENT'S ADDRESS

Prof Ryohei Takahashi
President, Kyushu University

Good morning, ladies and gentlemen. On behalf of Kyushu University, I would like to extend a hearty welcome to all of you who are attending the International Symposium on Biomagnetic Stimulation. It is an honor for us to have the opportunity to provide a forum for this rapidly developing interdisciplinary science which has important implications for the integrated aspect of both medical and engineering science.

In particular, my thanks go to the invited speakers from overseas, many of whom have traveled great distances in order to be with us here today. I understand that Professor Ueno has organized this international symposium as a satellite meeting of the World Congress on Medical Physics and Biomedical Engineering in Kyoto. I hope this meeting will serve to promote and stimulate interest in this new area among scientists and engineers in a wide range of fields. Additionally, it provides an opportunity for those of us who could not attend the Kyoto meeting to contribute to the discussion.

I confess, I'm surprised to know that our finger can be made to move without our intention when a strong pulsed magnetic field is generated in specific region of the brain. I believe that this scientific pioneering will open new horizons for the study of the functions of the brain and nervous system, leading to great contributions to the advancement of medical science and engineering and to the welfare of mankind.

Kyushu University has this year celebrated its 80th birthday. Since its foundation in 1911, our University has been involved in extensive research not only within Japan's shores but also on an international level, sharing cooperative research projects with foreign universities.

This area of Fukuoka has an extensive history of such international cooperation. Historically, it served as a gateway between Japan and the rest of Asia. Recently, in 1987, important historic ruins were discovered here in Fukuoka, called "Korokan." These ruins, constructed in the year 800, are believed to be the remains of a building that functioned as a guest house for foreign diplomats visiting the city. It is my hope that Kyushu University can serve a similar function, providing a center of international meeting and cultural and scientific exchange.

Thank you for your kind listening.

CONTENTS

THE HISTORY OF MAGNETIC NERVE STIMULATION AND ITS DEVELOPMENT AT THE UNIVERSITY OF SHEFFIELD

Ian L. Freeston

Department of Electronic and Electrical Engineering
University of Sheffield
Mappin Street
Sheffield S1 3JD
United Kingdom

This paper is about the history and development of magnetic nerve stimulation at the University of Sheffield, and its relationship with other work. It will deal with the early work in rather more detail than the subsequent development because other papers in this volume will treat these later topics in more detail. Before doing this, it is necessary to outline the principles involved.

Figure 1 shows a single turn coil producing a magnetic field; for simplicity, only one turn is shown, but in practice there are usually many turns. The magnetic field will penetrate materials, in particular materials that are weakly conducting, of which the human body is one. If the current in the coil producing the magnetic field changes with time, it will result in an electric field within the tissue. As the tissue is conducting, this will cause an electrical current to flow within the human body.

Figure 2 illustrates the application of this principle. For simplicity again, only one turn is shown, but the coil would normally be multi-turn. This current, if of appropriate magnitude, direction and duration, can cause stimulation of the nerve.

The advantage of this technique is that it is non-invasive and non-contacting; no direct contact with the arm is necessary. It can be relatively painless because no current flows through the surface, in contrast to electrical stimulation, which is the earlier, conventional technique. But how can currents of the right characteristics to cause stimulation be produced?

The effects of magnetic field stimulation have been known for many years. The English electrical engineer, S.P. Thompson [1], attempted brain stimulation in 1910. He conducted experiments on himself, and described magnetophosphenes. Earlier, D'Arsonval [2] in France had conducted similar experiments.

The interest at Sheffield in magnetic stimulation originated in about 1974, when Anthony Barker was a research student jointly supervised by myself and Professor Brian Brown. In his research project, he was looking at a method for measuring or evaluating the conduction velocity distribution of signals in nerves, and to do that it was necessary to stimulate the nerve electrically. A then-recent paper by Hallgren [3] had described the use of a magnetic field to stimulate nerves externally. This was investigated, and a slight effect obtained, but as it was a digression, it was not investigated in detail until after Dr. Barker had successfully completed his doctoral research.

The paper by Hallgren contained little quantitative information about response, and that was a difficulty with our earlier attempt to achieve stimulation. This was similar to the work of Bickford and Fremming [4] in 1965, whose work we were then unaware of; they had also reported a small effect, but had given no quantitative information about the response. In 1978,

Biomagnetic Stimulation, Edited by
S. Ueno, Plenum Press, New York, 1994

together with a research student, Michael Polson, Dr. Barker and I began a research project to study this effect and to see if we could develop magnetic nerve stimulation.

We considered ways in which it might be possible to obtain more effective magnetic stimulation. Figure 3, taken from the 1981 Ph.D. thesis of Polson [5], shows how we developed a magnetic stimulator by analogy with electric stimulation. This was done in about 1979, and we concentrated on peripheral nerves at that stage.

From experience, it was known that the voltage between two electrodes on the surface of the arm will produce stimulation. Therefore, at the site of a nerve, we can calculate the electric field necessary to produce stimulation using conventional electric stimulation. By analogy, in order to produce an induced electrical field at the same point of the same magnitude, we can calculate the current that would have to flow in a coil external to the arm. In this way, we were able to calculate the time-varying current needed, and to generate a specification for a magnetic stimulator which was subsequently constructed.

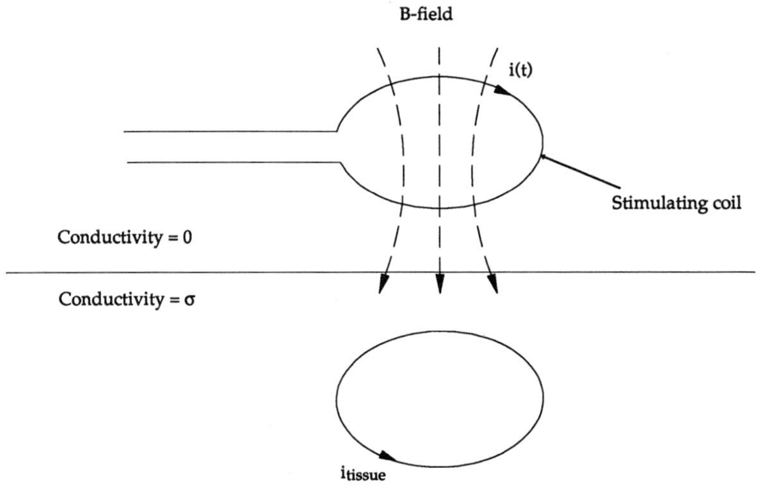

Figure 1. The time-varying field due to the stimulating coil can penetrate conducting tissue and induce currents.

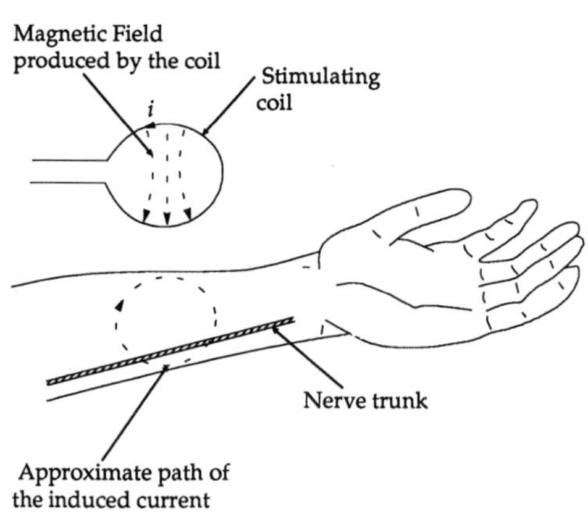

Figure 2. Induced currents of appropriate magnitude and time-dependence can be used to stimulate peripheral nerves.

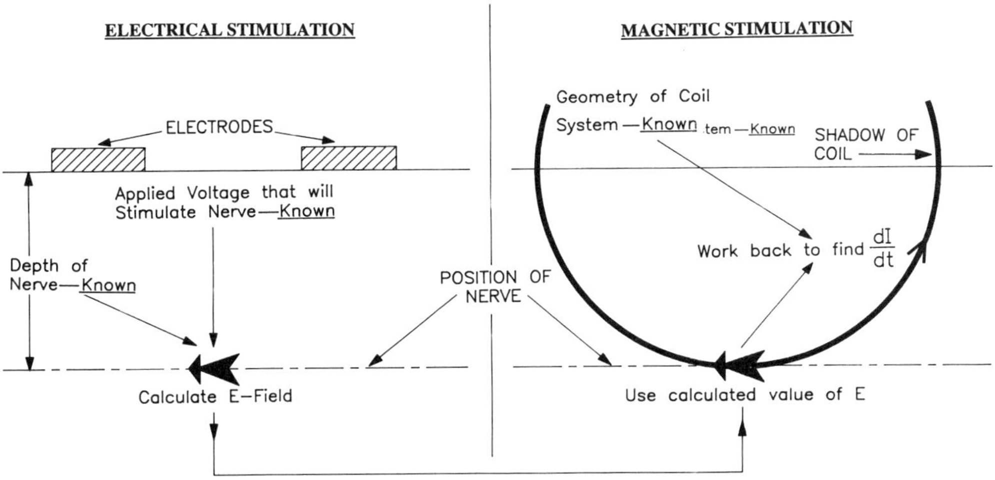

Figure 3. The electric field needed to stimulate a nerve is known from electrical stimulation. This value of field can be used to determine the magnetic field needed to stimulate the nerve.

The basic circuit of the stimulator is shown in Figure 4. It is a very simple circuit in which the capacitor C is charged from a power supply by closing switch S1. The capacitor is then discharged through the coil by closing the switch S2. The diode, D, and resistor R, shape the pulse appropriately. It is a very simple principle, but one that needs some engineering to implement because of the magnitudes of the currents and voltages involved. This basic circuit has been used in almost all stimulators subsequently developed.

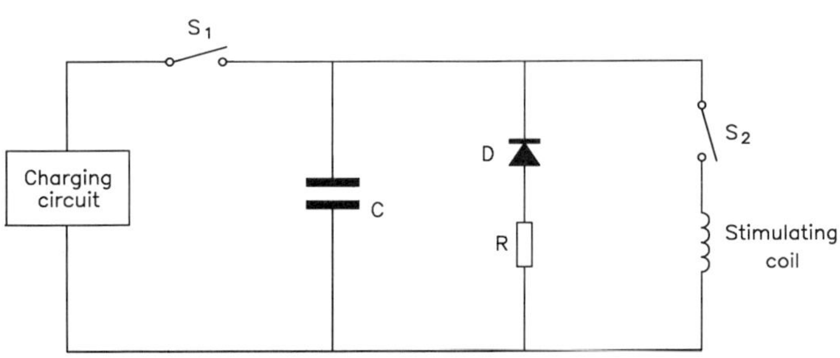

Figure 4. Basic circuit used to produce the magnetic field.

Figure 5 shows an early result obtained with the stimulator. It is an EMG measured on a muscle in the hand caused by stimulation in the forearm. The magnitude of the response is plotted against time; the stimulation takes place at time zero and there is an initial artefact. The solid line represents the response to magnetic stimulation. The comparison with electrical stimulation at the same site, here dotted, shows a very similar response, although with a different artefact. Figure 6 is a strength/response curve for the same experiment. The magnitude of the stimulus applied to the nerve is shown in terms of the capacitor voltage, and the threshold occurs at a voltage of about 200 volts. It rises very rapidly and then saturates. This is a fairly conventional response that has been reported several times. These results were published in 1981 [6].

Although these results clearly demonstrated stimulation of peripheral nerves using magnetic fields, one of the problems with the equipment we had designed was that it was underpowered, and we could not investigate the effect in more detail. A second research student,

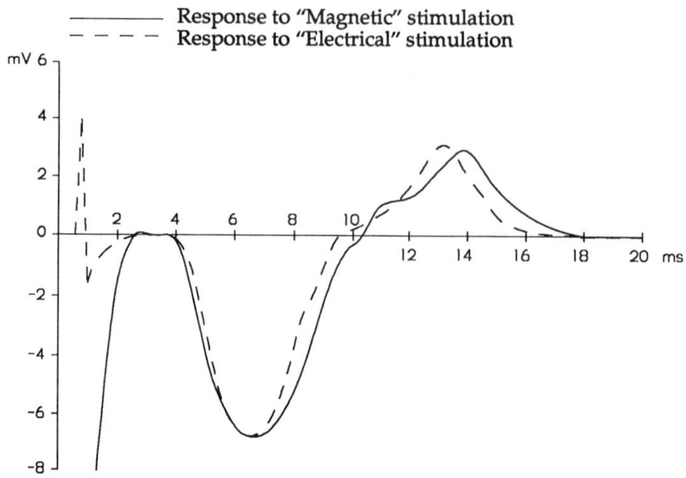

Figure 5. Comparison of responses to supramaximal electrical and magnetic stimuli.

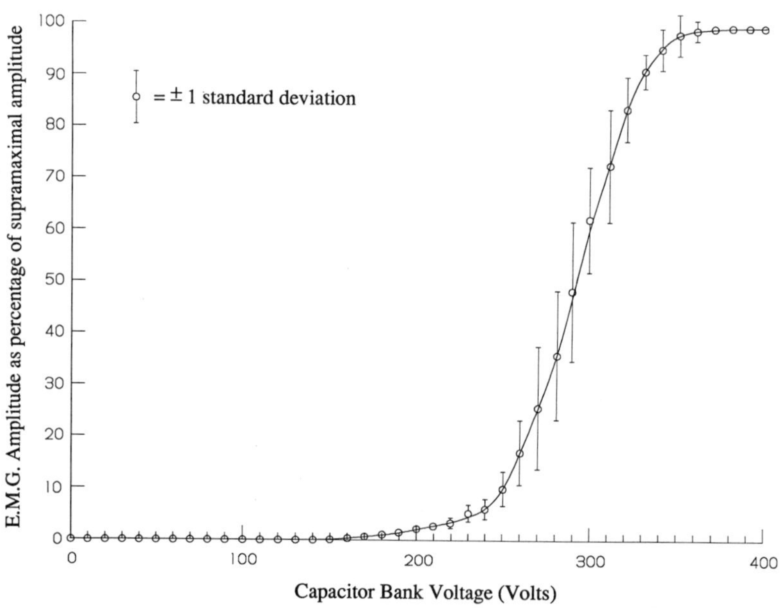

Figure 6. Typical strength response curve for magnetic stimulation.

Reza Jalinous, was asked to make a much more efficient stimulator. We developed the idea of a Figure of Merit for the stimulator, which is essentially related to the energy that is in the coil. With careful design, our next stimulator had a much higher efficiency; it was over 90 percent efficient, whereas our first stimulator had been under 50 percent. This efficiency was achieved through careful design and using non-electrolytic capacitors.

About the time that the stimulator was being constructed, we became aware of work being carried out by a research group in London using electrical stimulation of the cortex for various nervous disorders. We realised that it might be possible to use magnetic stimulation in this application, particularly as the stimulating current does not have to be injected through the skull, which is a relatively good insulator. This experiment was first carried out in February 1985 [8].

Other papers in this volume will describe in more detail the characteristics of magnetic stimulation of the cortex. Figure 7 shows a typical result obtained from stimulation of the cortex. It is clear that the technique enables the time of propagation of messages from the brain to be readily measured.

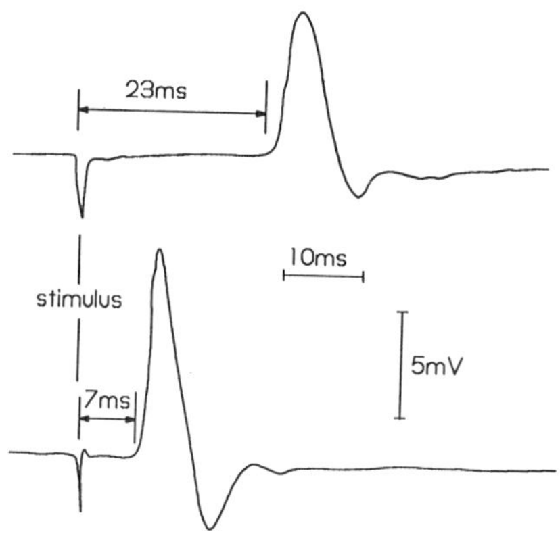

Figure 7. EMG recorded on the hand due to magnetic stimulation of the cortex (upper trace), and of the ulnar nerve at the elbow (lower trace). The signal takes 16 ms to travel from the brain to the elbow.

After this very successful experiment, we talked to a number of companies about the exploitation of the technology, including Novametric, Medelec and Digitimer in the United Kingdom and Cadwell in the United States, with the result that a number of companies are now manufacturing equipment. In 1985, we also made six machines for research groups, five of which were in the United Kingdom and one for Dr. Levy in the United States. It was already clear in 1985 that there were possibilities of applying the technology in a number of clinical applications. Another paper in this volume will consider the development of new methods, and of new areas of application.

There have also been technical developments since 1985, particularly the development of different coil geometries. The double coil, due to Professor Ueno, has been widely used [9]. Developments have taken place in the number of magnetic field pulses used. The work discussed so far relates only to single pulses, and some later work has used trains of pulses. The shape of the pulse and the timing could also have a significant effect on the response.

I do not intend to say anything on the question of safety of the technique. A number of speakers will draw attention to this very important issue and the need to ensure that a technique of this kind is made as safe as possible.

REFERENCES

1. S.P. Thompson, A physiological effect of an alternating magnetic field, *Proc. Roy. Soc.* 82: 396-8 (1910).
2. A. D'Arsonval, Dispotifs pour la mesure des courants alternatifs de toutes frequences, *C.R. Soc. Biol.* 3: 450-1 (1896).
3. R. Hallgren, Inductive neural stimulator, *in* : "IEEE Transactions," BME-20: 470-472 (1973).
4. R.G. Bickford and B.D. Fremming, Neuronal stimulation by pulsed magnetic fields in animals and man, *in:* "Digest of the Sixth International Conference on Medical Electronics and Biological Engineering," 112 (1965).
5. M.J.R. Polson, The stimulation of human peripheral nerve trunks by time-varying magnetic fields. Ph.D. Thesis, University of Sheffield, England (1981).

6. M.J.R. Polson, A.T. Barker and I.L.Freeston, Stimulation of nerve trunks with time-varying magnetic fields, *Med. Biol. Eng. Comp.* 20: 243-4 (1982).
7. I.L. Freeston, A.T. Barker, R. Jalinous and M.J.R. Polson, Nerve stimulation using magnetic fields, *in:* "Proceedings of the Sixth Annual Conference IEEE Engineering in Medicine and Biology Society" (1984).
8. A.T. Barker, I.L. Freeston and R. Jalinous, Non-invasive magnetic stimulation of the human motor cortex, *Lancet* 1: 1106-7 (1985).
9. S. Ueno, T. Tashiro, and K. Horada, Localized stimulation of neural tissue in the brain by means of a paired configuration of time-varying magnetic fields, *J. Appl. Phys.*, 64: 5862-4 (1988).

DISCUSSION

Chairman Prof. Åke Öberg: Thank you very much, Professor Freeston. Are there any questions?

Prof. S. Ueno: Thank you, Prof. Freeston. I have just one simple question: in developing the original equipment, which part was the most difficult? For example, capacitors? Or something else?

Prof. Freeston: The capacitors and the switching networks to handle the large currents involved.

In the very early work that we did with Polson, we used low-voltage, high-current circuits, which were very inefficient. We used current sharing between switches, and that gave problems. There is no doubt that using higher voltages as we did in later designs was a more technically sound solution.

Prof. Öberg: When you saw the potential in this electro-magnetic device for the first time, I suppose you approached industry solely on the basis of the industrial potential of this technique. What was the early reaction of industry when you approached it?

Prof. Freeston: Cautious interest. There was the possibility of some development taking place, but certainly in the United Kingdom, the market for specialised equipment is not a large one, and therefore I think U.K. companies had to ask themselves, "What is the international market for this?" For this reason, U.K. companies were, quite properly, a little cautious about getting into this market. And of course there were also safety conventions which they had to consider. In the end, a number of companies have chosen to enter the market, so I think it must have been seen as a distinctly realistic commercial opportunity.

Prof. S. Matsuoka: Could you tell me whether it is possible to give a double stimulation in the recovery cycle of the evoked potential?

Prof. Freeston: I don't know the answer to that, but I am sure that other people who are speaking today will be able to give an authoritative answer.

Prof. S. Hori: I'm deeply concerned about the safety of transcranial magnetic stimulation. What was the first response of the public in the United Kingdom? I think there is now a special committee, I don't know the exact name, possibly some larger committee in the United Kingdom that also deals with the safety of magnetic stimulation, the problems with magnetic stimulation, and that handled the alteration of the magnetic field's effect when MRI was invented in the United Kingdom.

Prof. Freeston: I am sure that safety issues will repeat themselves throughout the day, as safety is an important question. In terms of magnetic stimulation, no questions have yet been raised about its intrinsic safety. The dose level that one interprets in terms of exposure to magnetic fields is very low indeed, because the pulse repetition rate is low and the pulses last for very short times. Under such conditions, the only apparent physical effect that could occur is heating. If one does the calculation of how much power actually goes into tissue, per unit volume, it is very small, on the order of milliwatts, compared to the normal metabolic rate of a resting individual of about 70 to 100 watts. As yet, no other known effect has been substantiated.

Dr. Barker is a member of a committee in the U.K. that is looking at the effects of fields. Such committees have tended to concentrate their attention on the effects arising from power cables. Some technical and other journals have published studies that have found supposedly harmful effects. But the data is by no means wholly accepted by the scientific community as being statistically valid. I leave it to others in the audience, who know far more than I do about this subject, to address it further.

Prof. Öberg: Any other comments or questions? If not, then thank you very much, Professor Freeston.

MAGNETIC NERVE STIMULATION: PRINCIPLES, ADVANTAGES AND DISADVANTAGES

Anthony T. Barker

Department of Medical Physics and Clinical Engineering
Royal Hallamshire Hospital
Sheffield S10 2JF
United Kingdom

Chairman Prof. P.Å. Öberg: Professor Freeston defined the role of a supervisor as "transferring all problems to a Ph.D. student." Now we'll hear from the person who took care of all of the problems, Dr. Antony Barker, speaking on "Magnetic Nerve Stimulation: Principles, Advantages and Disadvantages."

Dr. A.T. Barker: Thank you, Mr. Chairman. Professor Ueno, distinguished colleagues, ladies and gentlemen, good morning.

I'd like in my talk this morning to give a broad overview of some of the practical problems and applications of magnetic stimulation. In the limited time available, I shan't go into very great detail in many areas. I've selected just one or two areas from within my overview, to go into a little more detail, because I think they are of some interest at the moment.

WHAT IS MAGNETIC NERVE STIMULATION?

Almost everybody in the audience, I'm sure, knows what magnetic stimulation is. I'd like to define it in a slightly different way from how engineers perceive it. I'd like to define it as a new solution to an old problem.

Ever since the work of Galvani and Volta in the 1790's, it's been known that nerves and muscles can be electrically stimulated. The technique of electrical nerve stimulation using either needle or surface electrodes is a widespread one today for diagnostic and therapeutic purposes, and is one with which we are all familiar. Magnetic stimulation is merely a different way of achieving the same effect, of getting electrical current into the tissue to cause nerve depolarization. We choose to use a magnetic field because it is a convenient vehicle. As we'll see later, it's got both advantages and disadvantages compared with electrical stimulation.

For those of you who are not engineers and who are unfamiliar with the technique, it essentially consists of a coil of wire through which a pulse of current is passed, as shown in Figure 1. The pulse of current generates a pulse of magnetic field which passes into the tissue and will induce currents in the body. If those currents are of the right amplitude and duration and the right geometric orientation to nerves of interest, they will stimulate those nerves, using the mechanism we are all familiar with of electrical membrane depolarization.

You have seen this photograph, Figure 2, already in Professor Freeston's talk. It was taken in 1910 and is, as far as I am aware, the first photograph of an attempt at magnetic brain stimulation. As you can see, it shows that standards have changed. In English universities we used to dress very formally when we did our experiments. Even in England, I'm afraid we no longer dress like that in our laboratories.

Biomagnetic Stimulation, Edited by
S. Ueno, Plenum Press, New York, 1994

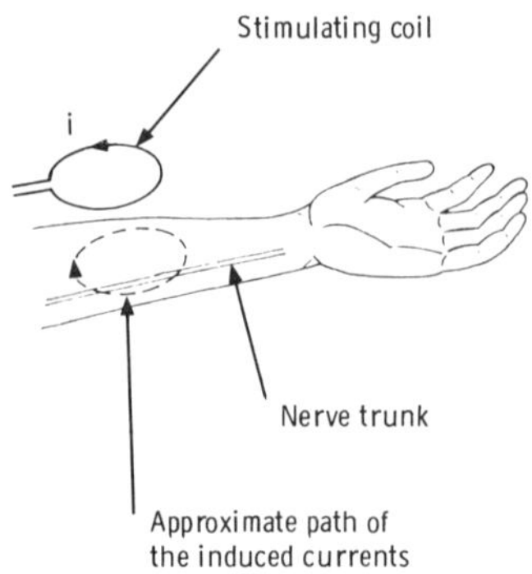

Figure 1. Schematic diagram of the basic principle of magnetic stimulation.

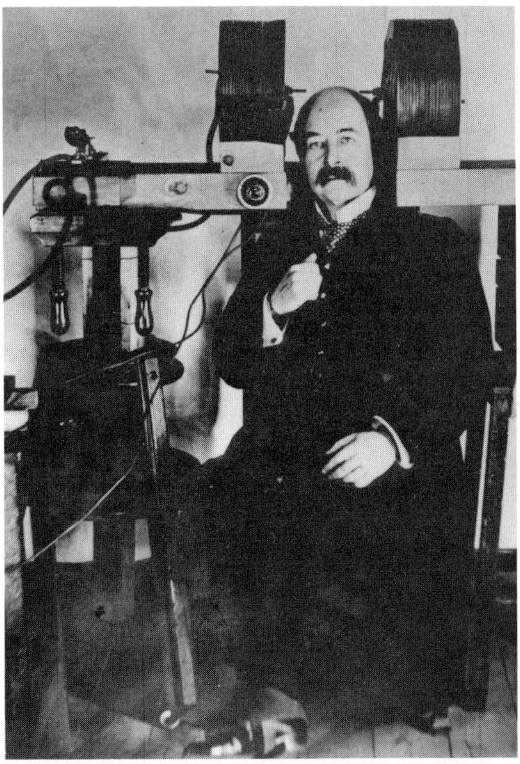

Figure 2. Silvanus P. Thompson trying to magnetically stimulate his brain. (London, 1910.)

It's been known for a long time that you could induce currents in tissue with magnetic fields. The principle is very simple. It's only relatively recently that it has achieved clinical applications for cortical and peripheral stimulation.

It took until 1985 for the first clinical magnetic stimulators to arrive. Pictured in Figure 3 is one of the six clinical magnetic stimulators, the first in the world, which were designed and built at the University of Sheffield.

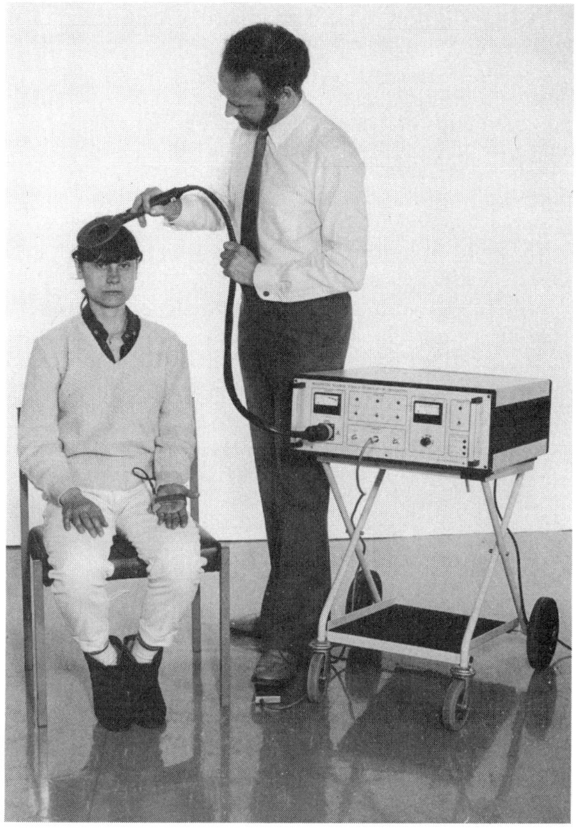

Figure 3. The first clinical magnetic nerve stimulator. (University of Sheffield, November 1985.)

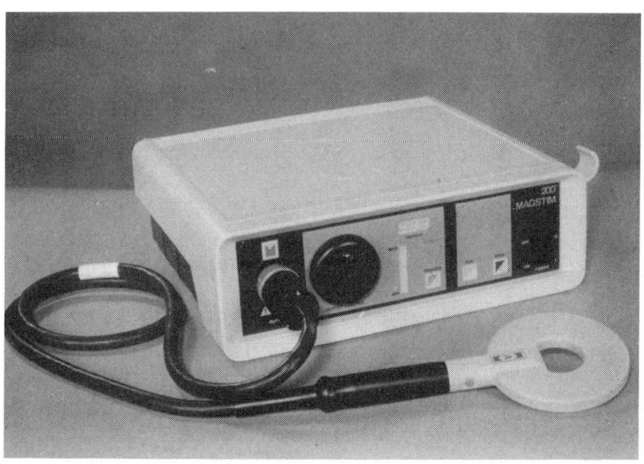

Figure 4. Commercial magnetic nerve stimulator made under licence to the University of Sheffield.

There has been considerable commercial interest in the technique since then. There are now commercial stimulators available from at least four manufacturers to my knowledge. I would estimate that there are now well over 500 commercial magnetic stimulators in use worldwide, the majority of them in Europe where the technique has got off to a rather faster start.

The stimulator pictured in Figure 4, just for the record, is a commercial stimulator made under license to the University of Sheffield, based on our original design. I understand that this model is the only machine presently licensed for sale in Japan by the Ministry of Health and Welfare.

DESIGN OF A SIMPLE MAGNETIC NERVE STIMULATOR

In our early research, we were able to show approximately the sort of magnetic fields that you need to generate (Figure 5). I'm going to say more about magnetic wave forms and strength of fields later on, but for now, if you will, just accept this as a typical magnetic wave form.

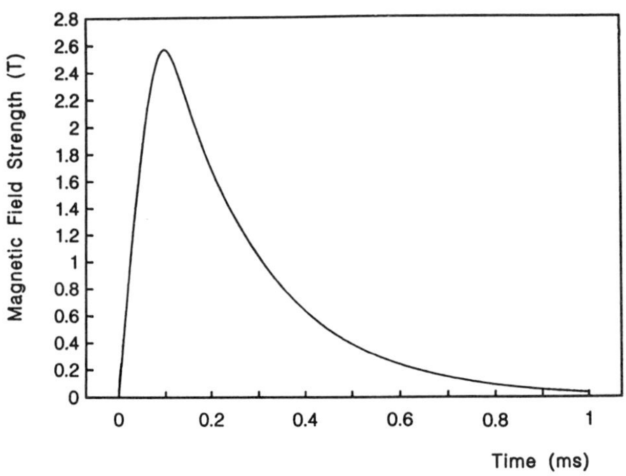

Figure 5. Typical magnetic stimulator output waveform.

The magnetic field is plotted vertically, with time plotted horizontally. The magnetic field starts at zero, rises rapidly to a peak and then decays back to zero over a period of time. It is during the rise time from zero to the peak that stimulation occurs. Typical rise times are of the order of 100 microseconds, typical magnetic fields are 1 to 2 tesla, depending on coil size and geometry. The rise time of the magnetic field does have an important bearing on the efficiency of stimulation, and I will mention more about that later.

What stimulates the nerve ultimately is current passing across its membrane, causing depolarization. Current induced in the tissue is proportional to the rate of change of magnetic field with respect to time. It's called "the first differential of the magnetic field with respect to time." You can find the equations in any basic physics textbook; it's a well-understood principle.

The wave form in Figure 6 is the first differential of the previous wave form. You can see it is a biphasic response. All magnetic stimulator currents induced in the tissue have at least two phases, and some have more. Stimulation usually occurs before the end or at the end of the first phase.

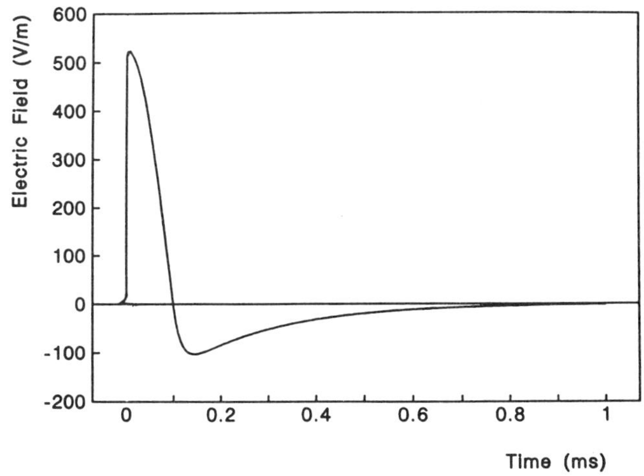

Figure 6. The rate of change of magnetic field with respect to time, to which induced current is proportional.

How does one achieve such large magnetic field pulses? While it was not very easy the first time around, like most things it's simple when you know how.

As shown in Figure 7, the easiest way of achieving the large pulses of current and magnetic field that are needed is with a capacitor-discharge system. A charging circuit is used to charge up an energy store made from storage capacitors of one type or another. These capacitors are then discharged through a high-voltage, high-current switch, into the coil to produce a current pulse, and, hence, a magnetic field pulse.

I use the terms "high-voltage" and "high-current" here advisedly. The engineering problems are not completely trivial; some stimulators produce output voltages of up to 4,000 volts, and some produce output currents up to 50,000 amps. Typical figures might be 3,000 volts and 7,000 amps: over 20 MVA. The pulse doesn't last for very long, but it's a jolly big pulse whilst it's there. Hence there are some interesting engineering problems in making magnetic stimulators.

Figure 7. Block diagram of a basic magnetic nerve stimulator.

I'd like now to look at the nuts and bolts of a magnetic stimulator, particularly for those of you who are not engineers, so that if you buy one you've got some vague idea of what's inside it.

You need a storage capacitor to store the energy inside a magnetic stimulator. They come in all shapes and sizes; pictured in Figure 8 is a typical one. They have an energy storage of anything from 500 joules to a couple of thousand joules, depending on the efficiency with which that energy is transmitted to the stimulating coil. The capacitors, certainly in the better stimulators, are usually purpose-built by capacitor manufacturers specifically for this application.

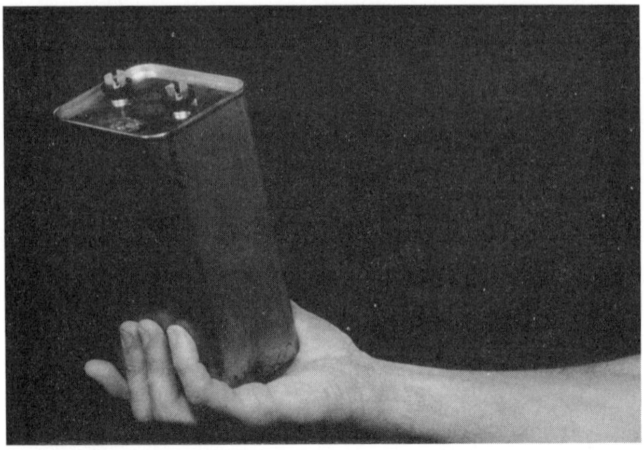

Figure 8. A typical capacitor used in a magnetic stimulator.

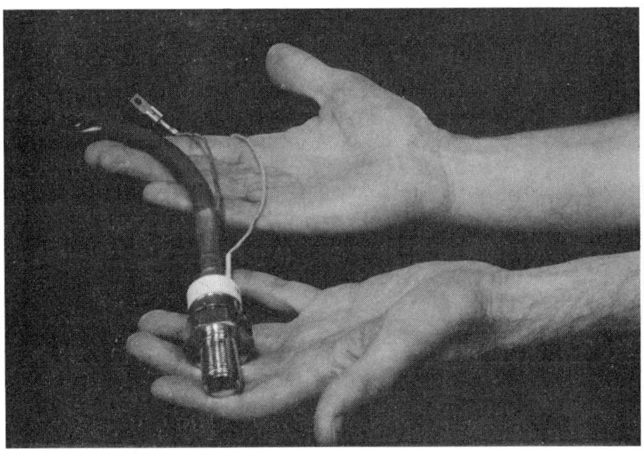

Figure 9. An example of a thyristor switch used in a magnetic stimulator.

You have to have a rather clever device to switch these high voltages and high currents. They are normally based on semi-conductor devices known as "thyristors" or "triacs." They all look somewhat different from each other.

Shown in Figure 9 is an unencapsulated version. Some of them come in little plastic blocks and look rather less interesting. You can get some idea of the current that's involved in these switching devices by the thickness of the wire that goes into the thyristor.

COIL SIZE, SHAPE AND ORIENTATION

But, of course, the business end of the magnetic stimulator, the bit you put on your patient, is the coil. And there are quite a few variables in coils. In particular, their size, their shape and how they are oriented relative to the subject.

Figure 10 shows what I call a large, circular coil. Designed and used primarily for cortical stimulation, it works very effectively, particularly for stimulating hand areas of the motor cortex.

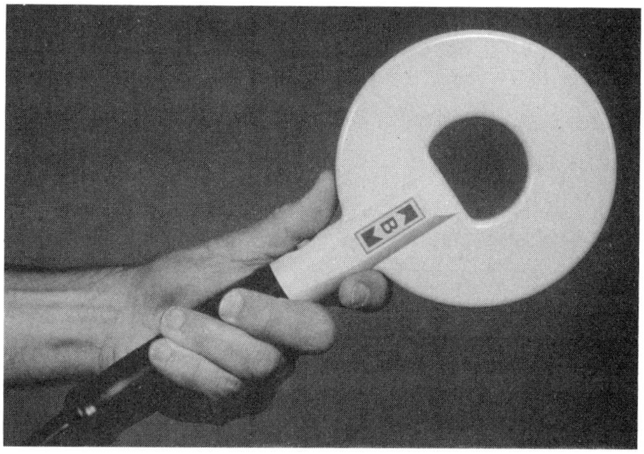

Figure 10. Large coil for cortical stimulation.

Such large coils, however, are not as useful in the periphery as they are cortically. You can perhaps see why that is in Figure 11. The coil is quite large compared with the anatomy at the wrist, for example. If you imagine the magnetic field coming out from all over the flat surface of the coil, some of it, quite clearly, will not pass through the tissue at all and will therefore have no useful effect. So it is important that you choose a size of coil that is fairly efficiently matched to the anatomy beneath that coil.

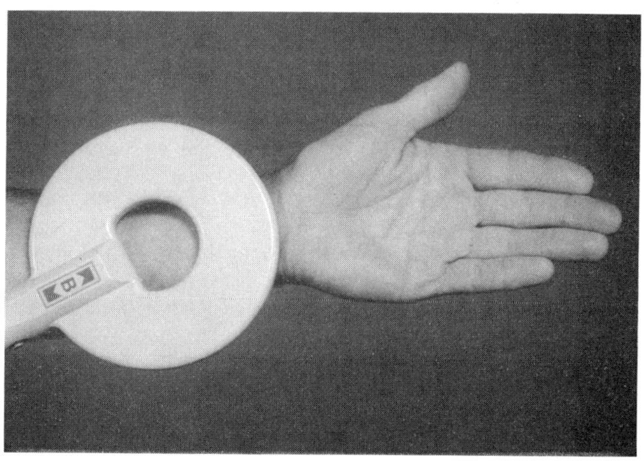

Figure 11. A large stimulating coil is less appropriate for stimulation in small peripheral areas.

So for peripheral work, you would perhaps use a smaller coil. There are, however, technical limitations on the size of coil that can be made. And indeed there are some technical disadvantages in very small coils, even if they can be made. Currently, the coil in Figure 12 is about the smallest coil that is routinely used. The mean coil diameter is somewhere just inside the blue line, and this is the loop under which the most effective stimulation will occur.

You can orientate a coil differently. The traditional way, the electrical engineer's way if you like, of putting a coil on an arm, is to lay it flat on the surface, because then the maximum amount of magnetic field which comes out of the flat surface of the coil passes into the tissue.

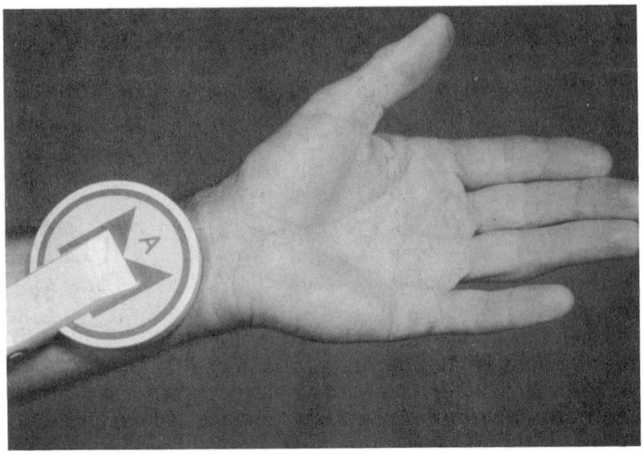

Figure 12. A small coil is often more useful for stimulation in the periphery.

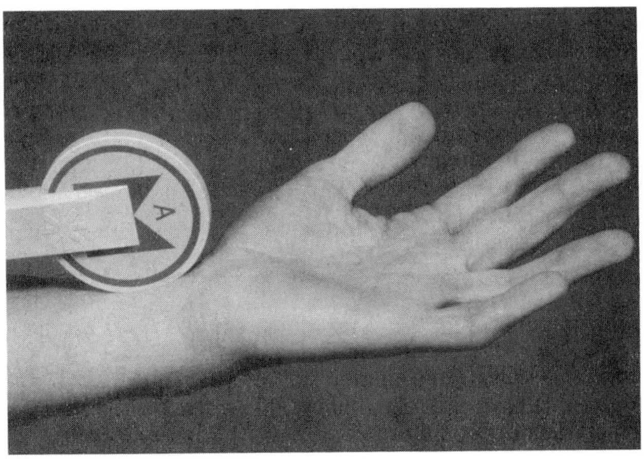

Figure 13. An alternative coil orientation.

It is possible, however, to stand the coil on edge as shown in Figure 13 and achieve stimulation. This geometry is very inefficient. Most of the generated magnetic field misses the tissue altogether. A small amount will come out of the lower edge, pass through the tissue and go back into the other side of the coil. There are some minor advantages in some anatomical situations in using this geometry, but it is a very inefficient configuration.

Figure 14 shows the only geometry of coil, other than the circular, that has been widely evaluated and has come into general use. It's the figure-of-eight or double-circular coil, about which I'm not going to say much because Professor Ueno, who's done much of the work on the development of this type of coil, is going to talk about it in some detail later. Suffice it to say that in some anatomical situations it has some advantages over a conventional circular coil.

It isn't, however, the solution to all problems. Again, if you try to use a double-coil of the size shown in Figure 14 in the periphery, most of the magnetic field will miss the tissue altogether and you'll be better off with a small circular coil for most applications.

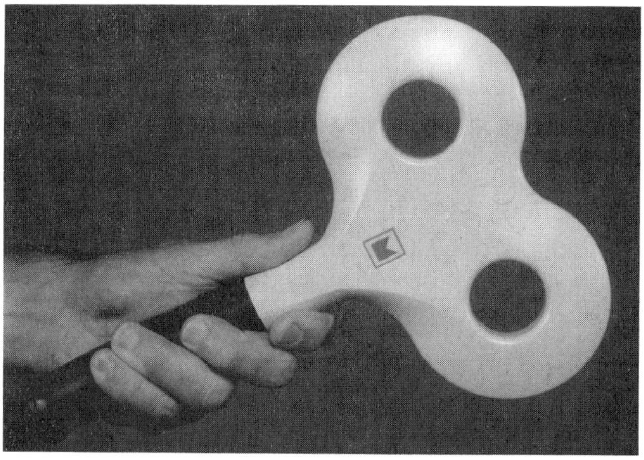

Figure 14. A figure-of-eight coil.

WHY USE MAGNETIC STIMULATION?

Why use magnetic stimulation? That is a very good question, because magnetic stimulation has a number of significant disadvantages when compared to electrical stimulation.

The equipment is relatively bulky and costly. The lightest commercial magnetic stimulator that I'm aware of weighs about 20 kg. (You saw a slide of it earlier.) And it'll set you back a good few thousand U.S. dollars, so they're not cheap. I can make an electrical nerve stimulator in a cigarette packet that'll run for weeks off a tiny internal battery and maybe cost $50. So, there's no doubt it's a more complex technology.

The pulse repetition rate of existing commercial magnetic stimulators is low, typically one stimulation every 2 or 3 seconds. There are variants on this: you can pair stimulators together so they'll fire two stimuli very close to each other, and perhaps Dr. Rothwell will speak about that as one of the applications of the technology later on this morning.

There are prototype experimental machines available that will stimulate at frequencies up to a few 10's of Hz for a short period of time before they overheat due to the electrical losses and coil heating. I shall talk a little bit more about those later. However, I think it's fair to say that for most E.M.G. recordings that you are likely to make, low repetition rate is not actually a problem. One stimulus is often sufficient to get the response that you require, and hence repetition rate is largely unimportant.

The site of magnetic stimulation is also somewhat less well-defined than with electrical stimulation, due to the size of current loops that are induced, particularly with large coils.

Magnetic stimulation does, however, have some significant advantages. One is that magnetic stimulation is able to penetrate bone structures such as the skull. The magnetic field is not attenuated in any way by the presence of the skull, unlike an electrical stimulus. If you try to stimulate the brain electrically by sticking two electrodes on top of the skull, you can achieve electrical brain stimulation only with difficulty. Because the skull has got high electrical resistance, the current is reluctant to pass through it. It is necessary to inject a very large current into the scalp to get a small percentage of it to pass through the skull into the brain, and this hurts.

Some techniques of electrical brain stimulation are less painful than others, but you typically have to put up to 1,000 volts between two electrodes on top of the scalp to get sufficient current penetration into the brain. Magnetic stimulation does not suffer from that quite significant disadvantage because the magnetic field goes through the bone without attenuation.

Which leads us to its second advantage: magnetic cortical stimulation really doesn't hurt at all, as hopefully you'll see later.

A third advantage of magnetic stimulation is that you can get greater depth of penetration than with electrical stimulation from surface electrodes. Hence, using magnetic stimulation, it is very easy to stimulate nerve structures such as deep peripheral nerves.

It is not necessary to make electrical contact with the subject because, obviously, there are no electrodes involved. Nor is mechanical contact needed, because the magnetic field will pass through air just as easily as it will pass through tissue. This means that we have a true non-contact method of stimulating nerves. This may have a number of applications, for example, in the stimulation of traumatized or damaged regions of tissue. When you wish to maintain sterility, you don't even need to touch your patient. And of course, the patient need not undress, which makes the technique quicker to use clinically, because you can stimulate through all clothing.

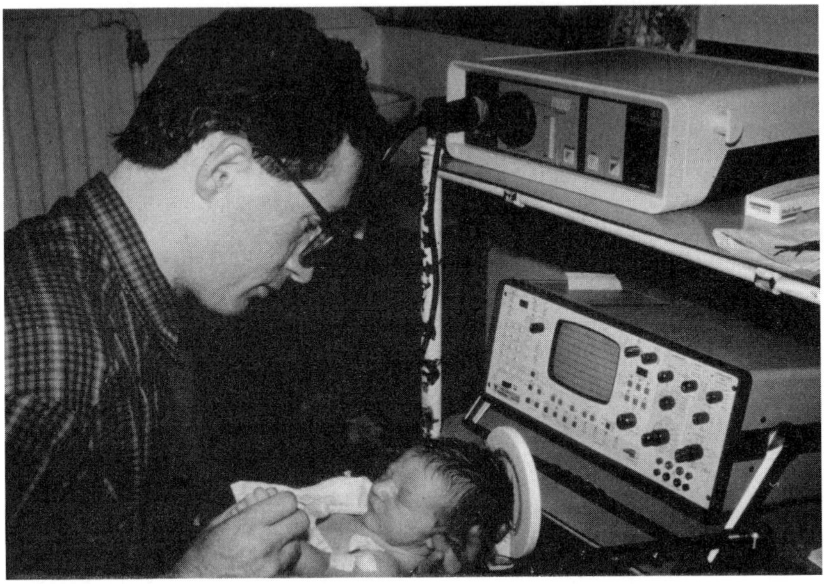

Figure 15. A neonate undergoing cortical magnetic stimulation whilst sleeping.

Figure 15 shows that it really doesn't hurt. You can use magnetic stimulation on anybody and it's pretty acceptable, even for neonates. Here we have a very young baby with a large circular coil placed at the back of his head, and recording electrodes on his little biceps. Babies will actually sleep through cortical magnetic stimulation.

It's easy to talk about magnetic stimulation -- and indeed I do talk for far too long about it -- but it's far more convincing to actually demonstrate the technique. Professor Ueno has volunteered to be my guinea pig for the second half of my demonstration.

Here is a commercial stimulator, coil and foot-switch. If I stand on the foot switch, it will fire once every 3 seconds or thereabouts, and you'll hear a slight click coming from the coil. I will talk more about that noise in a few minutes. It's very easy to demonstrate the technique, it clearly works. I can put the coil on any bit of my anatomy and I twitch every time the stimulator fires. As you can see, the stimulus goes through clothing, and it can't be that unpleasant because I can talk about it and do it at the same time. Obviously, EMG signals can be recorded using conventional technology in response to the mechanical twitch.

I hope I can show you that it's truly non-contact. I don't actually need to touch my arm at all, as I can still achieve stimulation with a couple of centimeters of air between the coil and my tissue. So, we have a truly non-contact method of stimulation.

You can also stimulate deep peripheral nerves without difficulty. For example, where the nerves of the arm go into the shoulder, up in the brachial plexus region, you can stimulate electrically, but it's not very pleasant. It's very easy, on the other hand, to stimulate that region magnetically. Again, I personally don't think it hurts at all, but I'm biased!

Now, Shoogo, I'll get you to sit down, if you would, and just hold up both your arms. A lot of the interest in magnetic stimulation, though it does have many other applications, is for cortical stimulation. And it's extremely easy to demonstrate with a coil of this size. If you put it centrally over the vertex, the mean diameter circle of the coil sits over the two regions of the motor cortex that control, primarily, the arms.

If I start at a relatively low level there will be no response. If I gradually turn the stimulus intensity up, you can perhaps see that both his hands are moving. It's very undramatic, really rather disappointing, but of course, there are large E.M.G.'s associated with those responses. If I put the coil in a different place, I can get his feet to move. Obviously, cortical stimulation must not hurt, otherwise Professor Ueno, who is a very sensible person, would not volunteer to have it done to him. Thank you, Shoogo.

I'm not going to talk in detail about the clinical applications of the technique, because there are far more qualified speakers, in particular Dr. Rothwell, who will be talking in detail on that subject later. However, I'd just like to very briefly give you a flavor for one of the possible ways that you can use a magnetic stimulator. Indeed, it was the first way that we thought of using a magnetic stimulator that could stimulate the human brain.

Put recording electrodes, for example, over abductor digiti minimi, the small muscle in the hand, or over any other muscle you're interested in. Stimulate cortically, record from that muscle and get an E.M.G. response. Re-position the coil in the spinal root area of the neck where the nerves of the arm come out of the spine. Stimulate there, record from the same muscle and get a second response.

These responses are usually a few millivolts in size, and hence you don't need to average. You can get strong responses from single stimuli.

The time difference between the latencies of the two responses is known as the "central-motor conduction time," and is due to pathways that are almost completely within the central motor system: the brain and the spine.

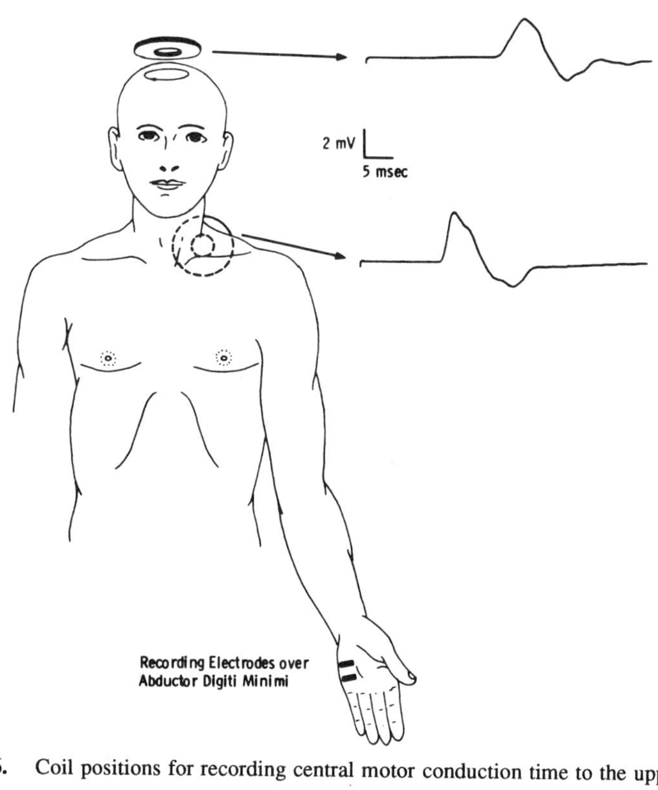

Figure 16. Coil positions for recording central motor conduction time to the upper limbs.

Figure 17 shows data from the first normal subjects studied with magnetic stimulation. Conduction time from stimulus to response is shown on the vertical axis in milliseconds. There are three groups of data. The first group shows data resulting from stimulating at the head, recording from a hand muscle. The average latency is 22 ms, or thereabouts, from head to hand. From neck to hand, the average conduction time is on the order of 13 ms. The third group shows conduction time from head to neck, a figure of 9 ms or thereabouts, which is almost entirely the central motor pathway conduction time.

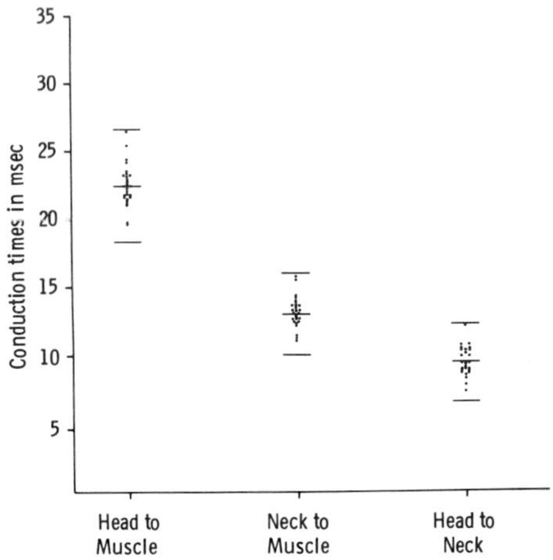

Figure 17. Conduction times recorded to ADM for normal subjects. Horizontal bars denote the mean and 99 percent confidence limits.

In Figure 18, we look at a group of patients. I've chosen Multiple Sclerosis (MS) because it was the group that we first investigated with the technique. The data you see between the small horizontal bars is exactly the same data as on the previous slide, the normal population. The dots to their right are patients who had already been diagnosed as having Multiple Sclerosis and, as a very crude measure of the severity of the disease, we divided the patients into two groups: those who were mobile, who could walk; and those who were chair-bound, and were not able to walk, indicated as MSm for "mobile" and MSc for "chair-bound."

Looking at the data for the MSm group you can see that, from head to hand, there was prolonged latency. In those who were chair-bound there is an increase of latency. From shoulder to hand the patients are within the normal range, with the exception of one patient who had an ulnar nerve lesion at the elbow.

Subtracting shoulder-to-hand latencies from head-to-hand latencies gives head-to-neck conduction times. And you can see that every patient in the Multiple Sclerosis group falls outside the normal range apart from a single subject.

One can also see that many of the patients who are chair-bound have much longer latencies than those who were mobile. Some of these latencies, given that the normal figure is about 8 to 10 ms, are grossly prolonged, several tens of milliseconds, which is just what you'd expect from a central demyelinating disease.

MAGNETIC STIMULATOR WAVEFORMS

I'd like now to go back to technical matters and talk about the different waveforms that you can choose to get from a magnetic stimulator. There are quite a lot of technical constraints on the waveforms that can be produced, but there are some options.

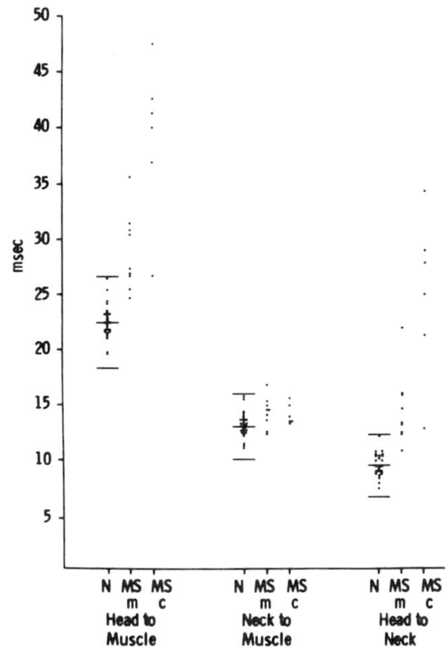

Figure 18. Data from a group of MS patients, recorded from the abductor digiti minimi muscle.

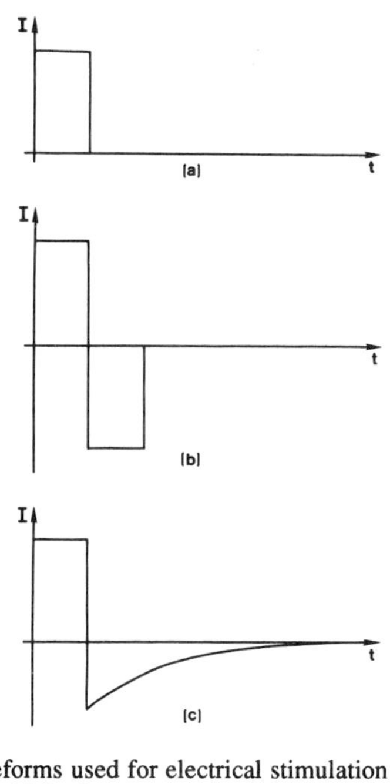

Figure 19. Typical current waveforms used for electrical stimulation : a.) monophasic, b.) symmetrical biphasic, and c.) experimental decay biphasic.

Figure 19 shows schematic representations of the sort of waveforms that might come out of an electrical nerve stimulator using conventional electrical technology. You can have what I would call a monophasic rectangular current pulse, as shown in Figure 19a; a symmetrical bi-phasic or charge-balanced current pulse, as shown in "b"; or, you can have a rectangular pulse with charge balancing occurring in an exponential phase, as at "c". Most stimulators that don't specify their output waveform approximate to waveform "c" in which no net charge is put into the tissue, a large, short-duration pulse is produced and then a small-amplitude, long-duration tail comes subsequently.

The sort of wave forms that you can produce with a magnetic stimulator are limited by technological aspects and can roughly be catagorized by comparison with the four wave forms that you see in Figure 20. A magnetic stimulator cannot produce the first wave form shown in Figure 19. It is technically impossible with realistic technology to produce a magnetically induced, uni-directional current pulse in the tissue, as this would require the coil current to rise to a peak and stay there indefinitely. So all realistic magnetic stimulators, because of the basic physics of induction, produce balanced pulses with no net charge injected into the tissue.

But you can choose waveforms to some extent. Pictured in Figure 20 at "a" is the nearest waveform you can practically produce to the symmetrical bi-phasic waveform that was pictured in Figure 19 at "a".

Instead of collapsing the magnetic field quickly, you can allow it to ring for some period of time, as shown in Fig. 20b. There are disadvantages to the ringing waveform. It exposes the patient to more magnetic field and more current for a longer period of time and in general, does not achieve more powerful stimulation. It also has the disadvantage that you do not know when, exactly, on this time axis you are stimulating the nerve. And, obviously, when you stimulate is important for latency measurements

Most commercial machines, and most prototype machines in laboratories, produce waveforms along the lines of the two shown in "c" and "d" of Figure 20. The difference between "c" and "d" is merely a difference of time scale. Pulse "c" does not last as long; it is a squeezed-up-in-time version of the pulse "d".

How do you decide which of those two sorts of wave forms -- a short-duration or a long-duration impulse -- you would like your stimulator to produce?

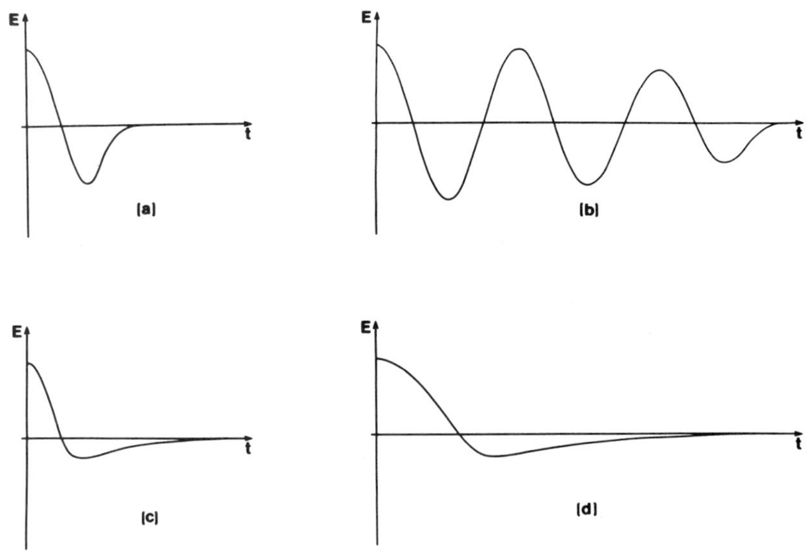

Figure 20. Practical induced current waveforms which can be produced with magnetic stimulation: a.) overdamped, b.) underdamped, c.) critically damped and d.) critically damped with long magnetic field risetime.

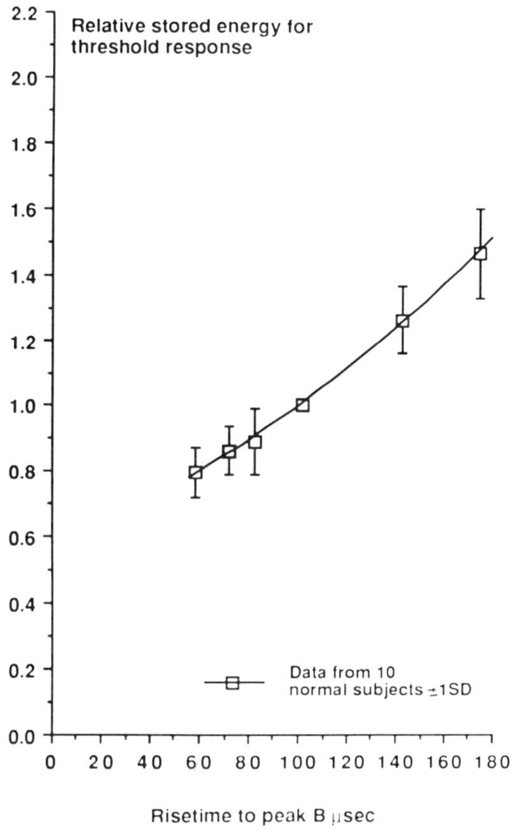

Figure 21. Cortical stimulation thresholds. The relative stored energy within the stimulator required to achieve threshold cortical stimulation versus magnetic field risetime.

The ability of a stimulator to stimulate is affected by the choice of waveforms. Figure 21 shows averaged data from 10 normal subjects. On the vertical axis, we're plotting a measure of the energy stored in a stimulator whose waveform is alterable and can be adjusted between the two bottom wave forms, "c" and "d" in Figure 20, and also intermediate wave forms. The coil is on top of the head and we are measuring the energy that is required from the stimulator to achieve a threshold response. A measure, if you like, of the efficiency of the overall stimulator/stimulation process.

On the horizontal axis we are plotting the time that the magnetic field takes to rise to its peak, in other words, a measure of the time scale of the stimulating pulse. You can clearly see that you need less stored energy to stimulate with shorter rise time magnetic fields than you do with longer rise time magnetic fields, a factor of 2:1 over the rise time range of about 60 to 180 microseconds that we have considered.

Why does the stored energy required to achieve stimulation depend on magnetic field rise time? The answer is twofold:

First, the charge induced in the tissue is proportional to the square root of energy in the coil. That doesn't depend on the rise time of the stimulator. You could get that energy in slowly or quickly and the total charge induced in the tissue would, roughly speaking, be the same in both cases.

However, as I'm sure many of you are aware, nerve membranes act as what I call "lossy integrators." By that I mean that in a very basic way you can model the nerve membrane, for example, at a node in a myelinated fiber, as a resistor and a capacitor in parallel with each other. The object of neural stimulation is to depolarize that membrane, that is, to change the voltage across the capacitor. If you inject charge slowly into the membrane it will tend to leak away through that resistor as quickly as you're putting it in, and you'll hardly change the membrane capacitor voltage at all. If you put the same amount of charge in quickly, you will charge the capacitor before the charge has time to leak away through the resistor. This is analogous to electrical stimulation, where short duration electrical stimuli are also more efficient than long duration stimuli, as shown in the traditional strength-duration curve.

This can be clearly demonstrated in the laboratory, as we see in Figure 22. Using the same axes as we had before, showing rise time to the peak of the magnetic field versus stored energy in the stimulator, you can cause the induced current to flow into a capacitor and resistor modeled to look like a nerve, with different values of resistor to simulate the nerve membrane time constants. Figure 22 shows different simulated neural membrane time constants, the more steeply sloping curve here has a short time constant, the longer time constant has the more horizontal line. If the time constant were infinite, the line would be horizontal and stimulation efficiency would not depend on the stimulator rise time.

Now, I'm sure you've already realized that you can superimpose the subject data of Figure 21 onto the family of curves as shown in Figure 22.

In Figure 23, the dotted curves are the curves of the capacitor and resistors modeling a nerve membrane. The solid line running down the center is the data from our 10 normal subjects. You can see two things from this figure. First, that a resistor and a capacitor offer a reasonable model of a nerve membrane, because the real data fits with the theoretical data. Second, you can measure a nerve membrane time constant with an experiment such as this, which it is not possible to do (certainly in the brain non-invasively) in any other way.

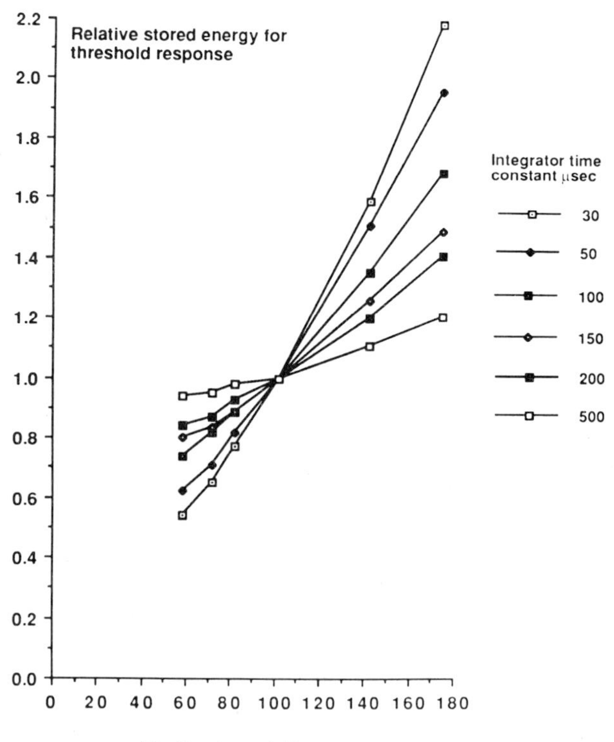

Figure 22. dB/dt integrated using different time constants. The relative stored energy required to give an arbitrary voltage change on a capacitor, with a variable shunt resistance to model membrane time constant, versus field risetime.

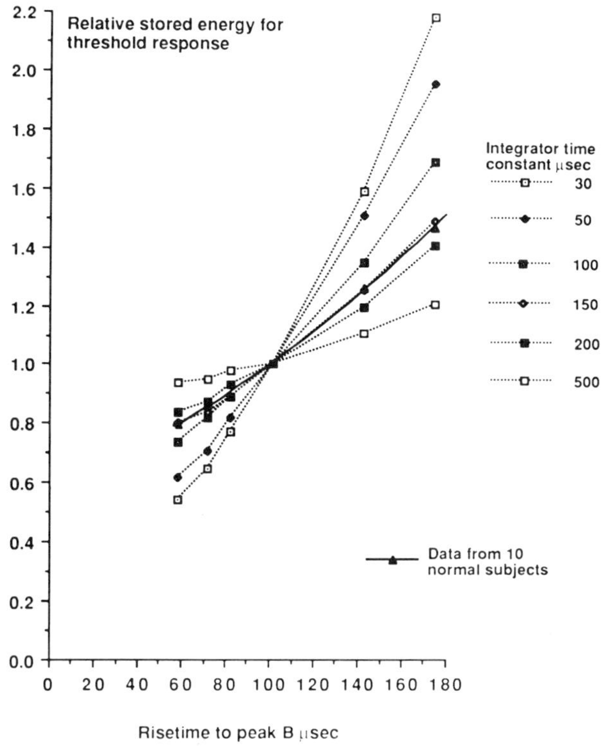

Figure 23. Cortical stimulation thresholds. The data of figures 21 and 22 superimposed.

The answer for normal subjects comes out to a time constant of about 150 microseconds, a figure not quantified before. Its measurement may have diagnostic implications and also basic physiological implications, shedding light on which structures are being stimulated.

MEASUREMENT OF STIMULATOR OUTPUT

Another area I've chosen to talk about is how to measure the output of a given coil/stimu
lator combination in laboratory conditions. A lot of nonsense is talked about how to measure the output from a magnetic stimulator. Manufacturers will thrust glossy brochures in your hands saying their stimulator is the most powerful, because its dB/dt or peak magnetic field is bigger than anybody else's. But it isn't that simple. For example, a small coil will give a larger magnetic field when plugged into the same stimulator. That does not mean it stimulates more efficiently. It is not as simple as peak magnetic field or rate of change of magnetic field alone.

The effective output from a combination of a magnetic stimulator and a coil depends on a number of factors. Some are anatomical, over which one has no control from an engineering point of view and I shan't mention. Factors of an engineering nature include the peak magnetic field, the rise time of the magnetic field, and the geometry of the coil and how well it couples to the tissue.

If you really want to measure the efficiency of a magnetic nerve stimulator, it's quite complicated. The best method I can suggest is to measure E, the induced electric field underneath the coil at the depth that you're interested in. Then it is necessary to integrate that electric field with the time constant that the human nerve actually has, probably about 150 microseconds. Any other approach will give you a rather meaningless, simplified estimate of the output of the stimulator.

Using a dipole immersed in saline, E is not technically difficult to measure. It can be made from three wires twisted together and you measure the voltage difference between the tips of two wires, spaced about 1 cm apart, compared with the reference wire in the centre. This is standard technology for measuring electric field. If the dipole output is then integrated with the appropriate time constant, you will have a measure of stimulator output.

It's got a rather awkward unit. We would suggest that it is called the "Et_{150}" measure. "E" because it is a measure of electric field, "t" to show that it's an integral with time and 150 because that's the time constant that we would recommend for the integral. The units are rather inconvenient: volt-seconds per meter. But if you produce that single measure, you can then at least make an attempt at comparing the relative efficiencies of your stimulators.

An easier way of measuring would be to get all of the manufacturers to bring their stimulators to your laboratory. Having decided which bit of the body you wish to stimulate, you can put a coil from each stimulator on that bit, and measure the percentage of the output of the stimulator to get a threshold response. This is rather crude and unsophisticated, but it will give a good approximation for the particular application you have in mind. Having said that, it is also true that most stimulators now have sufficient power for most applications.

Is cortical magnetic stimulation safe? A questioner raised that subject a few minutes ago, and Professor Freeston dealt with it briefly. I haven't got time to go into all the factors involved, but suffice it to say that the summary answer, as far as we're aware, is yes. There are, however, a number of areas one can consider, and the literature has looked in detail at most of them. I just want to emphasize three areas in particular.

Kindling: For those of you who are not familiar with the term, Kindling is the generation of an electrogenic focus or an epileptic-type center within the brain. Kindling was demonstrated many years ago using electrical stimulation and implanted electrodes within an animal brain. However, those experiments have failed to show Kindling phenomena being generated at stimulus frequencies below 10 Hz, no matter how long the animal preparation was stimulated, often for many days. So, the existing generation of magnetic stimulators, with low repetition rates, would appear to pose no real risk of causing Kindling. If magnetic stimulators with frequencies above a few Hz become commercially available, however, perhaps that subject would need readdressing, particularly if they are going to be used on normal subjects. It is likely that fast-repetition rate cortical stimulation may cause seizures in some normals.

Mechanical and electrical design of the stimulator: Magnetic stimulators have a lot of energy in them. If it is put directly into a patient, it could kill him or her, just as if you stick your fingers inappropriately into a television set. Therefore, it is important that the stimulator and the coil should meet the (usually very stringent) safety standards of the country in which you are working. Don't buy a stimulator that doesn't meet those safety standards. Also I would suggest that the coil needs particularly careful attention because it's the bit closest to the patient. It needs to be constructed such that, if the winding inside the coil did fail, then the housing of the coil would protect the patient from the mechanical and electrical energy involved. Again, the better magnetic stimulator makers have obviously taken this into account and, as far as I'm aware, no injury has ever been caused to a patient by a magnetic stimulator.

Acoustic output: As you heard when I demonstrated magnetic stimulation, a click is produced when the stimulator is fired due to slight movements of the turns of wire inside the coil. It has been suggested in one paper that this click might cause damage to one's hearing. We have measured the output from one stimulator with two coils -- by coincidence it is the machine that is currently available in Japan, and which I demonstrated here today. We have found that the acoustic output from the coils is a long way below the nationally regulated thresholds for acoustic output in the United Kingdom. We have chosen to compare our measurements with U.K. regulations, but I would imagine similar standards apply in Japan.

This conclusion is supported by the fact that there have been no clinical reports of damage due to the clicks from magnetic stimulator coils. It's actually somewhat quieter than snapping your fingers very close to your ear. The machines have now been used widely for several years, and there have been no reports of hearing damage. Perhaps, however, manufacturers should be encouraged to quote the acoustic output from their coils, not because it's a problem, but simply to reassure the user that it isn't. If users are still concerned, either for their own or for their patients' safety, the use of earplugs can provide on the order of 20 dB attenuation of the sound.

CONCLUSION

To summarize, there are four main points in my talk:

--Magnetic stimulation is pain-free and well-tolerated by subjects ranging from neonates to adults.

--Because it's pain-free, you can do repeat measurements and long-term studies on patients. Patients will come back time and time again and have the tests done on them, because it is so acceptable. And you can get normal volunteers; I'll do it, anybody will do it from your own laboratories, because it doesn't hurt.

--It's quick to use. The patient need not undress and you don't need to make mechanical and electrical contact with the subject.

--You can stimulate deep peripheral nerves and the motor cortex with ease.

Finally, for all these reasons and several others that I haven't time to go into, I think it's reasonable to say that magnetic stimulation is likely to have a wide range of future clinical and research applications.

Thank you, ladies and gentlemen.

SELECTED BIBLIOGRAPHY

Barker, A.T., 1991, An introduction to the basic principles of magnetic nerve stimulation, *J Clin Neurophysiol* 8:26-37.

Barker, A.T., Freeston, I.L., Garnham, C.W.,1990, Measurement of cortical and peripheral neural membrane time constant in man using magnetic nerve stimulation, *J Physiol* 423:66.

Barker, A.T., Freeston, I.L., Jalinous, R., Jarratt, J.A., 1987, Magnetic stimulation of the human brain and peripheral nervous system: an introduction and the results of an initial clinical evaluation, *Neurosurgery* 20:100-9.

Barker, A.T., Garnham, C.W., Freeston, I.L., 1991, Magnetic nerve stimulation -- the effect of waveform on efficiency, determination of neural membrane time constants and the measurement of stimulator output, *EEG & Clin Neurophysiol* , Supplement 43.

Barker, A.T., Jalinous, R., Freeston, I.L., 1985, Non-invasive stimulation of the human motor cortex, *Lancet* 1:1106-7.

Barker, A.T., Stevens, J.C., 1991, Measurement of the acoustic output from two magnetic stimulator coils, *J Physiol* 438:301.

Boyd, S.G., Kandler, R.H., Stevens, W.R., 1991, A comparison of noise levels produced by different magnetic stimulators, *J Physiol* 438:368.

Counter, S.A., Borg, E., Lofqvist, B., Brismar, T., 1990, Hearing loss from the acoustic artefact of the coil used in extracranial magnetic stimulation, *Neurology* 40:1 159-62.

Eyre, J.A., Miller, S., Ramesk, V., 1991, Constancy of central conduction delays during development in man: investigation of motor and somatosensory pathways, *J Physiol* 434:441-52.

Geddes, L.A., 1991, History of magnetic stimulation of the nervous system, *J Clin Neurophysiol* 8:3-9.

Kandler, R.H., Jarratt, J.A., Davies-Jones, G.A.B., Gumpert, E.J.W., Venables, G.S., Sagar, H.J., Zeman, A., 1991, The role of magnetic stimulation as a quantifier of motor disability in patients with Multiple Sclerosis, *J Neurol Sciences* 106:31-34.

Lovsund, P., Öberg, P.Å., Nilsson, S.E.G., 1980, Magnetophosphenes: a quantitative analysis of thresholds, *Med Biol Eng Comput* 18: 326-34.

Polson, M.J.R., Barker, A.T., Freeston, I.L., 1982, Stimulation of nerve trunks with time-varying magnetic fields, *Med Biol Eng Comp* 20:243-4.

Thompson, S.P., 1910, A physiological effect of an alternating magnetic field, *Proc R Soc Lond (Biol)* 82:396-8.

Ueno, S., Tashiro, T., Harada, K., 1988, Localized stimulation of neural tissue in the brain by means of a paired configuration of time-varying magnetic fields, *J Appl Phys* 64:5862-4.

DISCUSSION

Chairman Prof. P.Å. Öberg: Thank you , Dr. Barker, for an excellent presentation. You certainly solved the problems that were transferred to you in an excellent way. I particularly liked your demonstration. That way of presenting scientific results was very common in the old days, though we've tended to forget in the last 50 or 100 years. An actual demonstration was an excellent way of telling us how magnetic stimulation works. There is maybe time for one single question.

Dr. M. Fujiki: Thank you, Dr. Barker. I was surprised that the data you presented today showed very small stimulus artefact. It is difficult to record short latency potential from the spinal cord without stimulus artefact. Could you tell me how to record the smaller stimulus artefact, please?

Dr. Barker: Thank you. A very good question. Those of us who are familiar with magnetic stimulation use recording techniques that those of us who are not, tend not to use. They are what I would call good electro-physiological practices.

Electro-physiology recording equipment is so good these days, that you can often get away without skin preparation and with unscreened leads to your recording equipment and not have technical problems. Because of the large time-varying magnetic-field inducing currents all around the patient and around the recording amplifiers, there are a number of fairly obvious things that you can do to minimize stimulus artefact. But you cannot get rid of it completely, just as you can't when using electrical stimulation.

The key things to do are to make sure your amplifier is some distance away from the stimulating coil; don't put the amplifier next to the magnetic stimulator. And very important, have a screened lead from the recording equipment to the patient, to minimize the amount of current induced in the preamplifier. Also, use a good design of recording amplifier, by which I mean, one that does not recover slowly from overloads. Some recover very rapidly and it is usually the case, even with very good technique, that your recording amplifier will be overloaded for the duration of the stimulator pulse. If your recording amplifier has good recovery characteristics, it will recover from that overload very quickly and then be active in the period during which you are interested in recording a signal.

These are all guides. Good electrode preparation also minimizes artefact from magnetic stimulation. They are not complete solutions to the problem, but they go a long way toward helping. If you are still having problems with your preamplifier, you can develop your own electronic switch which disconnects the preamplifier for the fraction of a second that the magnetic stimulator is activated, thus stopping it from receiving an overload signal. We published a brief paper describing how to do this, it's fairly simple technology to make your own, essentially, fast switches in the electrode leads.

Prof. Öberg: Thank you very much, Dr. Barker.

FOCAL AND VECTORIAL MAGNETIC STIMULATION OF THE HUMAN BRAIN

Shoogo Ueno

Biomagnetics and Biosystems Laboratory
Department of Computer Science
and Communication Engineering
Kyushu University
Fukuoka 812
Japan

INTRODUCTION

Magnetic nerve stimulation has been studied for almost a century. Successful magnetic stimulation of the human brain was reported by Barker et al. in 1985. Since then, many further developments have taken place. We have developed a method of focal and vectorial magnetic stimulation of the human brain. In this chapter, we describe our early and more recent work in that area. We also propose a model of neural excitation elicited by magnetic brain stimulation; the model explains our observation that the optimal direction of stimulating current vectors for neural excitation reflects both the functional and anatomical organization of neural fibers in the brain.

DEVELOPMENT OF MAGNETIC NERVE STIMULATION

Magnetic stimulation of excitable tissues has been studied by several investigators [1-5]. Among them, Maass and Asa proposed a transformer type of stimulation in which a nerve bundle was threaded through a core as the secondary winding [4]. They demonstrated that the flux change in the core could be used to excite nerves. Öberg proposed an airgap type of stimulation, in which a nerve bundle was exposed to alternating magnetic fields [5]. Induced eddy currents in the membrane tissues could be expected to stimulate nerves. Although these types of magnetic nerve stimulation had been experimentally demonstrated, the underlying nerve excitation processes following magnetic stimulation were not yet understood [6].

We carried out an experiment to measure action potentials of lobster giant axons under time-varying magnetic fields [7, 8]. The axon membrane was excited by galvanic stimulation, and the action potential was recorded intercellularly with micro-electrodes. During the propagation of the action potential along the axon, alternating or pulsed magnetic fields were applied across the middle of the axon in order to study whether or not the magnetic fields have any effect on parameters such as conduction velocity and refractory period of the nerve fiber and amplitude, duration and shape of the action potentials.

The results obtained from the lobster experiments suggest that nerve excitation by magnetic field influence is mediated via the induction of eddy current in the tissue surrounding the nerve. The current density induced depends on geometrical factors as well as the resistivity of the tissue in which the current flows. In other words, for nerve excitation, the microscopic eddy currents that flow inside the microstructures of the axon membrane are not very impor-

Biomagnetic Stimulation, Edited by
S. Ueno, Plenum Press, New York, 1994

tant. What is important are the macroscopic eddy currents that flow along the nerve axon and in the tissues surrounding the nerve, as they contribute to the depolarization of the membrane.

After the study of single nerve axons, we proposed a new type of magnetic nerve stimulation [9]. An insulated magnetic core is implanted in the body with a nerve bundle positioned on the core aperture. The nerve can be stimulated by the eddy currents that flow in the body fluids around the core when the magnetic flux in the core is changed. There is no interlinkage between the core and the nerves. Therefore, this method is a good example for verifying that the nervous system responds to time-varying magnetic fields through eddy currents induced in the body.

BEGINNINGS OF MAGNETIC BRAIN STIMULATION

As mentioned above, magnetic stimulation of the human brain was first reported by Barker et al. [10-12]. Parameters of muscular potentials, such as conduction velocity, latency

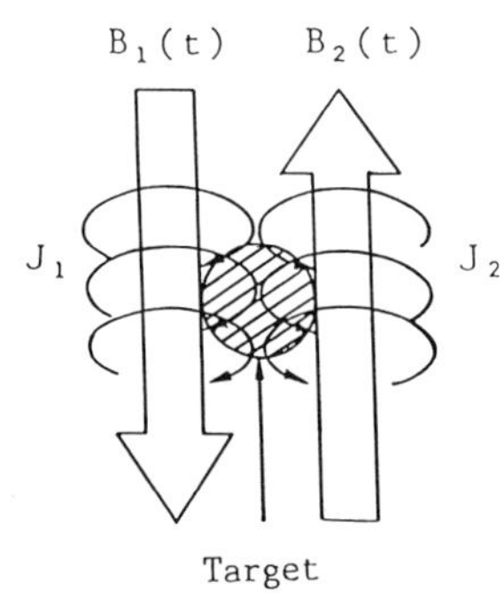

Figure 1. Principle of localized stimulation. Magnetic fields B_1 (t) and B_2 (t) are applied in opposite directions around the target area.

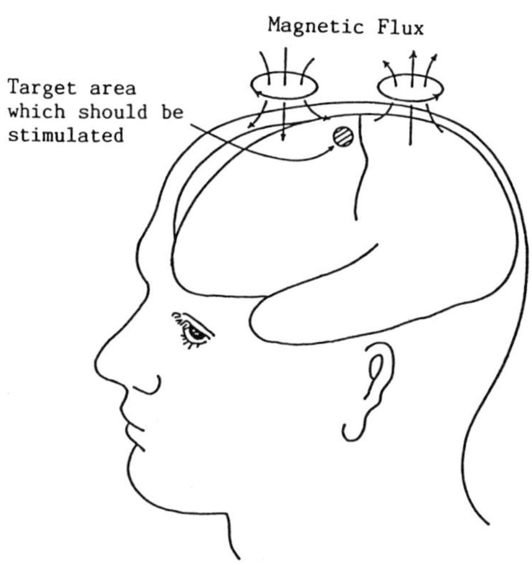

Figure 2. Principle of localized stimulation of the cortex. A pair of coils is placed outside the head so that the magnetic fields pass through the head in opposite directions around the target.

and amplitudes were studied by a number of researchers by recording the electromyographic (EMG) responses to stimulation of the motor cortex [13-17]. Single coils were used for these studies. Current pulses were passed through a single coil placed outside the head, and eddy currents induced in the head by the pulsed magnetic fields stimulated the brain. With this method, however, broad areas of the brain are stimulated simultaneously. We developed a method of localized magnetic stimulation of the human cortex [18].

The basic idea is to concentrate induced eddy currents locally in the vicinity of a target by a pair of opposing pulsed magnetic fields which can be produced by a figure-of-eight coil. Using this method, we were able to stimulate the motor cortex of the human brain within a 5 mm resolution [19, 20]. Since the concentrated eddy currents beneath the intersection of the figure-of-eight coil flow in a direction parallel to the tangent of both circular coils, vectorial stimulation can be achieved. The neural fibers can be excited easily when the fibers are stimulated by the eddy currents that flow parallel to the nerve fibers.

Based upon this principle, we obtained functional maps related to the hand and foot areas [21-23]. We have observed that an optimal direction of stimulating currents for neural excitation exists in each functional area in the cortex. We have also observed that the functional maps of the cortex vary with the orientation of the stimulating current. To explain the mechanism that is responsible for producing this anisotropic response to brain stimulation, we developed a model of neural excitation elicited by magnetic stimulation [24].

PRINCIPLE OF LOCALIZED STIMULATION

Figures 1 and 2 show the principle of localized stimulation of the brain [18]. The basic idea is to concentrate induced eddy currents in the target area by a pair of time-varying magnetic fields. A pair of coils is positioned outside the head so that time-varying magnetic fields B1(t) and B2(t) pass through the head in opposite directions around the target. The induced eddy currents J1 and J2 are expected to flow together. This convergence of eddy currents acts to raise the current densities in the target area, where depolarization of neural tissues can be caused.

This method is a different application of the same principle we previously proposed to heat brain tumors [25, 26].

CALCULATION OF EDDY-CURRENT DISTRIBUTIONS

The human head is modeled by a spherical conductor of a dimension 12.0 units in diameter with a uniform conductivity. The surface of the sphere is made up of 58 planes, as shown in Figure 3.

A pair of coils of a dimension 2.0 units in diameter is positioned outside the head at X = 0.0, Y = +2.0 or -2.0, and Z = 6.0 units. These coils generate magnetic fields in opposite

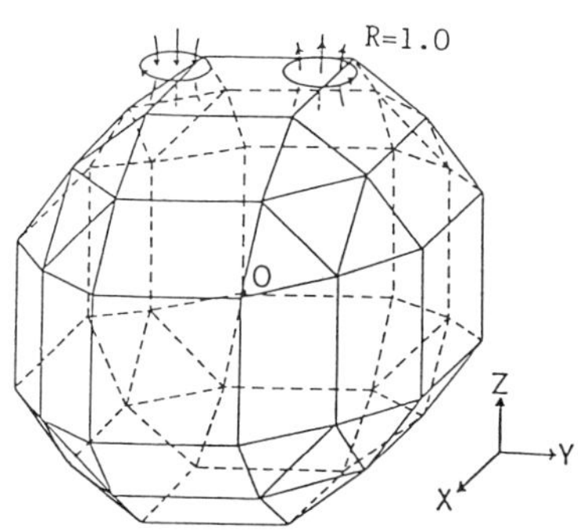

Figure 3. A quasi-spherical volume conductor and paired-coil configuration. The surface of the sphere is divided into 58 planes.

directions when currents are applied to the coils in opposite directions.

The distributions of eddy currents in the conductor can be calculated using a finite element method. The quasi-spherical conductor with 58 sides is divided into 5,504 small elements of a tetrahedron.

The calculated results are shown in Figures 4 and 5. Figure 4 shows current distributions induced in cross sections at Z = 5.125 and 3.5 units. The current vectors form two vortexes which flow together at the target between coils, so that eddy currents are concentrated in the target area.

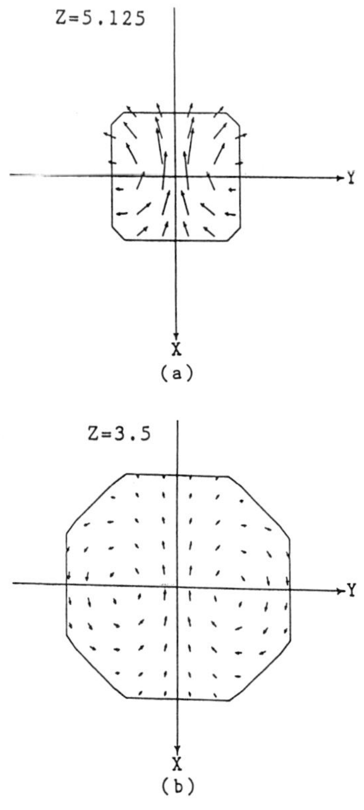

Figure 4. Distributions of eddy currents flow in a cross section of the X-Y plane produced by a pair of one-turn coils.

Figure 5 shows current distributions induced in X - Z cross sections at Y = 0.5 units and in Y - Z cross section at X = 2.5 units. The current density decreases with the increase in depth from the surface. It is understood that the current density is higher only at the target near the surface of the head.

We also investigated the effect of distance between coils on the current density induced at the target. For simplicity, a cubical model of a dimension 10.0 units on a side with a uniform conductivity is used for this study as shown in Figure 6. A pair of coils of a dimension 2.0 units in diameter is positioned at X = 0.0, Y = +2.0 or -2.0, and Z = 6.0 units.

Spatial distributions of eddy currents induced in (X - Y) cross section at Z = 4.5 units are shown in Figure 7. The eddy currents flow together along the X-axis.

Let us calculate eddy current densities on the Y-axis in Figure 7, assuming that the resistivity p = 1/o = 3 ohm-m, current in the one-turn coils I = 1.0 kA, frequency f = 1.0 kHz and 1 unit = 0.01 m.

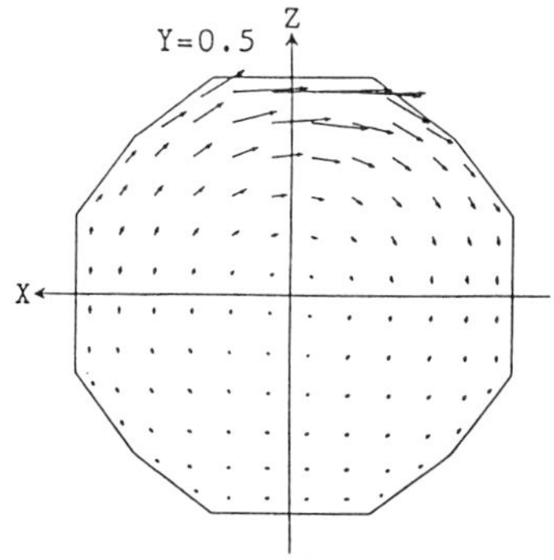

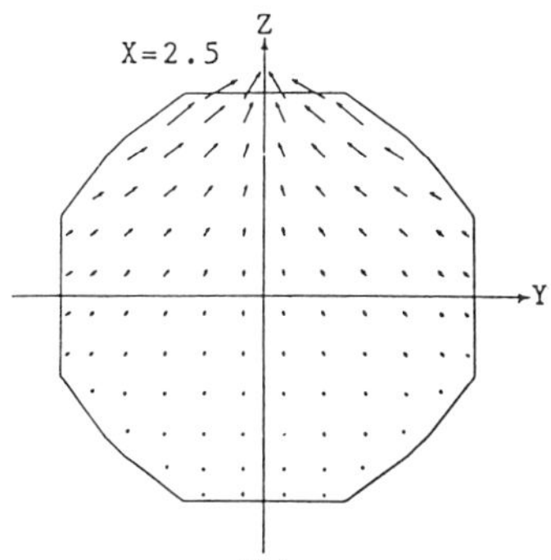

Figure 5. Distributions of eddy currents in a cross section of (a) the X-Z plane and (b) the Y-Z plane produced by a pair of one-turn coils.

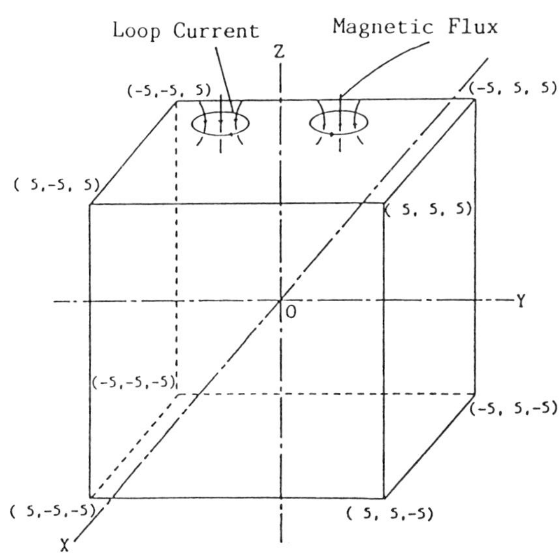

Figure 6. Cubical model and paired-coil configuration.

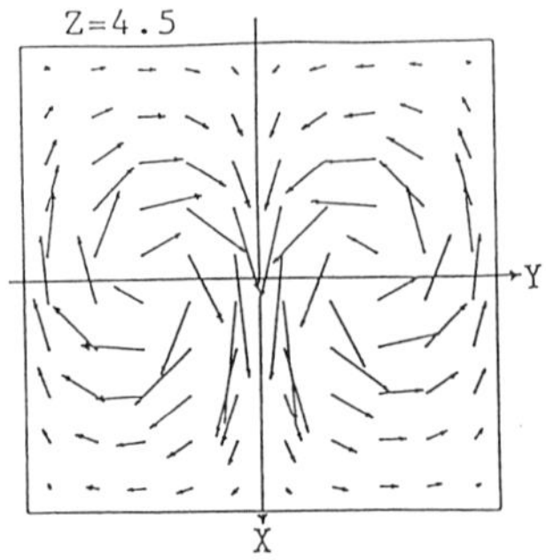

Figure 7. Distributions of eddy currents in a cross section of the X-Y plane produced by a pair of one-turn coils. Coils are positioned at a distance of 1.0 unit from the surface.

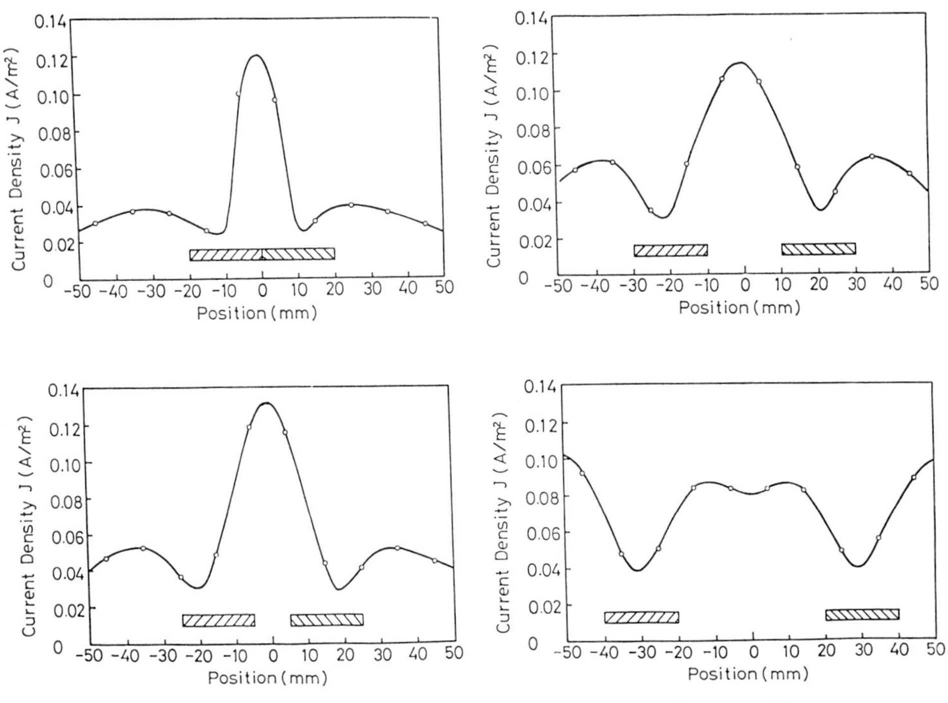

Figure 8. Changes of current density with the distance between coils.

Figure 8 shows the results. The current density at the target between two coils, marked with a hatched pattern, makes a peak that is higher by two to three times than the current densities at non-target regions. The ratio of the peak value at the target to the values at non-targets increases with a decrease in distance between coils, and the ratio is highest when the coil distance is zero.

EXPERIMENTAL METHOD

Based on a computer simulation, we have developed a method to stimulate localized areas of the cortex using a figure-of-eight coil, as shown in Figure 9. A figure-of-eight coil is positioned outside the head so that time-varying magnetic fields pass through the head in opposite directions around a target that is to be stimulated. The induced eddy currents are expected to flow in the same direction. This convergence of eddy currents acts to raise the current densities in the target where depolarization of neural tissue can be caused. The target is a point just under the intersection of the figure-of-eight coil. Since the concentrated eddy currents at the target flow in the direction parallel to the tangent of both circular coils, vectorial stimulation can be achieved. The various parts in the motor cortex are stimulated, and the corresponding muscular action potentials are recorded by electromyograms (EMGs) from the hand and foot muscles.

The induced current density decreases with depth, ie., with the distance from the coils to the target. In determining functional maps, the depth for the threshold intensity is important. In all stimulation trials, both the distance of the intercept from the head and the intensity of the current waveform in the coils are maintained constant for all orientations and positions in each experiment.

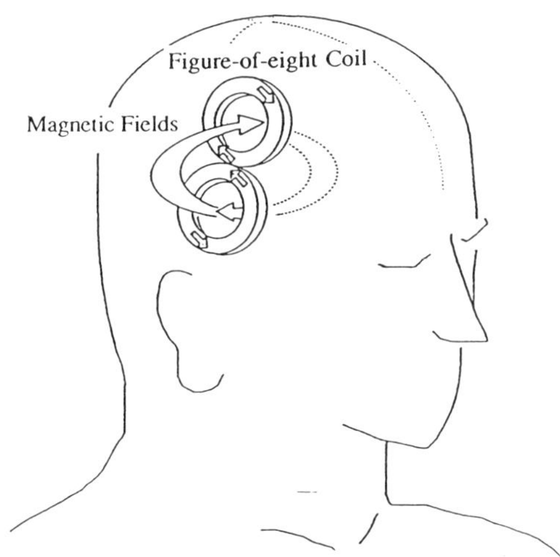

Figure 9. Principle of localized stimulation of the brain. A figure-of-eight coil is positioned outside the head so that magnetic fields pass through the head in opposite directions around a target. The target is located beneath the cross-point of the figure-of-eight coil.

RESULTS OF TARGET STIMULATION

Magnetic stimulation of the human motor cortex related to the hand area was achieved. We defined the coordinates on the surface of the head. The Y-axis was chosen so as to coincide with the line along the central sulcus. The Y-axis passes a point 20 mm posterior to Cz of the 10-20 system and reaches to the center of the zygomatic arch. The X-axis is oriented in a direction perpendicular to the Y-axis, and crosses the Y-axis at 40 percent lateral from the point 20 mm posterior to Cz. (The distance between the point 20 mm posterior to Cz and the center of the zygomatic arch is 100 percent.)

Definite EMG responses to magnetic brain stimulation were obtained. Figure 10 shows the results in the case of the thenar muscle. A figure-eight coil was placed at a point A (X = 15.0 mm, Y = 0.0 mm), called the "thenar point". Definite EMG response was obtained at the thenar point, or point A. No responses were obtained at points 5 mm away: points B, C,

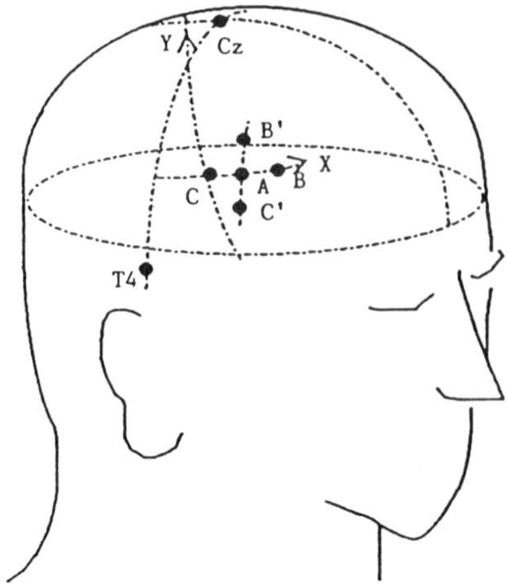

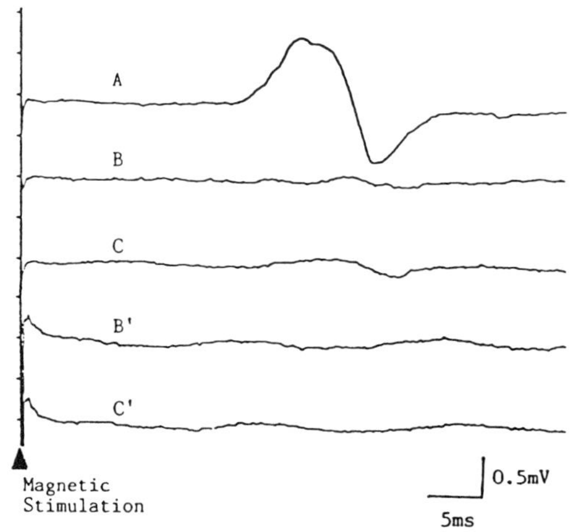

Figure 10. EMG responses to magnetic brain stimulation. A: Thenar point stimulation. B: Stimulation at a point 5 mm posterior to the thenar point. C: Stimulation at a point 5 mm anterior to the thenar point. B': Stimulation at a point 5 mm above the thenar point. C': Stimulation at a point 5 mm below the thenar point. (© 1990 IEEE)

B' or C'. In other words, the selective stimulation of the human motor cortex was realized within a 5-mm resolution.

If we plot the weak signals as well as the strong signals of EMG recordings from the thenar muscle, we can obtain a contour map of the localized functional area of the motor cortex.

Figure 11 shows the functional localization of the motor cortex related to the thenar muscle in a subject if his hand muscles are totally relaxed. The effect of muscular tension on spatial distributions of functional organization of the cortex was also investigated.

Figure 12 shows the distribution of functional organization of the cortex related to the thenar muscle if the same subject maintained a 10 percent maximum contraction of the muscle. Compared with the results in Figure 11, the area of the activated neural population in the cortex is spread out, and a part of the region is shifted into the somatosensory area.

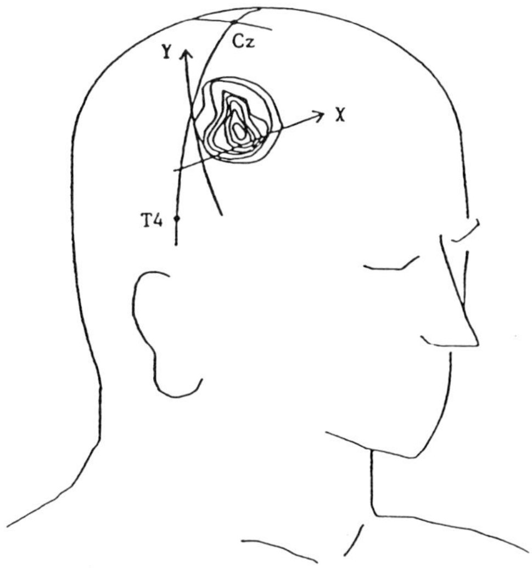

Figure 11. Functional localization of the human motor cortex related to the thenar muscle in one subject with the muscle totally relaxed. The equipotential lines which show peak-to-peak values of EMG responses to magnetic brain stimulation are quantized by 0.4 mV from 0 to 3.2 mV.

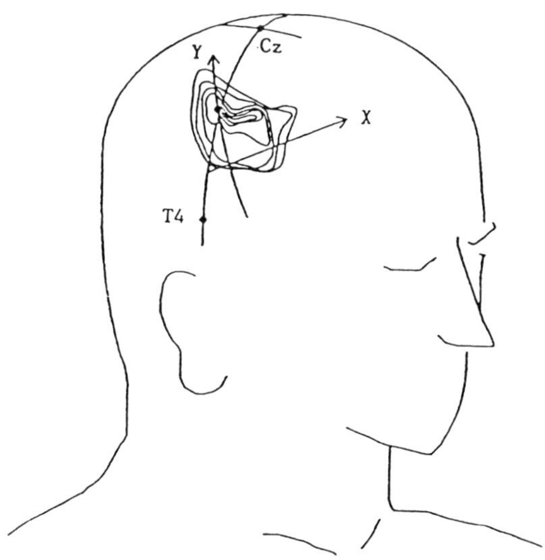

Figure 12. Functional localization of the human motor cortex related to the thenar muscle in the subject maintaining a 10 percent maximum contraction of the muscle. The equipotential lines which show peak-to-peak values of EMG responses to magnetic brain stimulation are quantized by 2.0 mV from 0 to 12.0 mV.

It has been reported that both latency and amplitude of the EMG responses to magnetic stimulation of the motor cortex are dependent on voluntary preactivation of the muscle [15, 17]. It has been difficult, however, to study the spatial properties of activated neural population in the cortex using a single coil technique. Magnetic brain stimulation using a figure-of-eight coil has enabled us to achieve more exact stimulation. The experimental results showing that stimulation of the somatosensory area in the cortex produces finger movements suggest the existence of neural connections from the somatosensory area to the motor cortex.

RESULTS OF VECTORIAL MAGNETIC STIMULATION

It is important to know whether or not an optimal axis exists in the cortex for brain stimulation. Let us say that axis 1 is oriented in a direction perpendicular to the midline which passes Cz and T4 of the 10-20 system as shown in Figure 13. By rotating a figure-of-eight coil at the cross point of the coil, we stimulated the thenar point in the motor cortex.

The peak-to-peak values of EMG signals that responded to the oriented brain stimulation were plotted on each axis as shown in Figure 14.

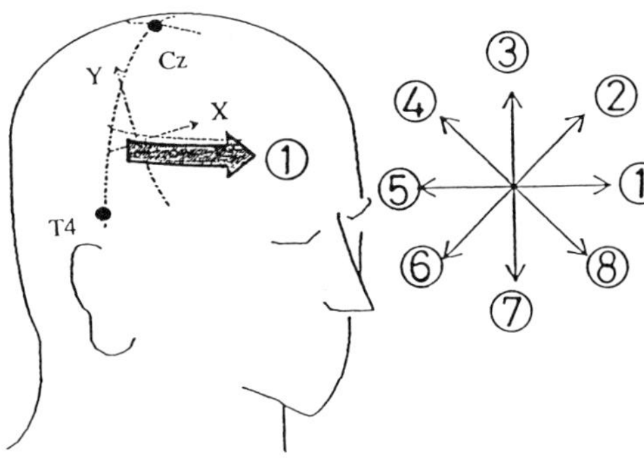

Figure 13. Distribution of stimulating currents. The axis 1 orients perpendicular to the mid-line that passes Cz and T4 of the 10-20 system. (© 1990 IEEE)

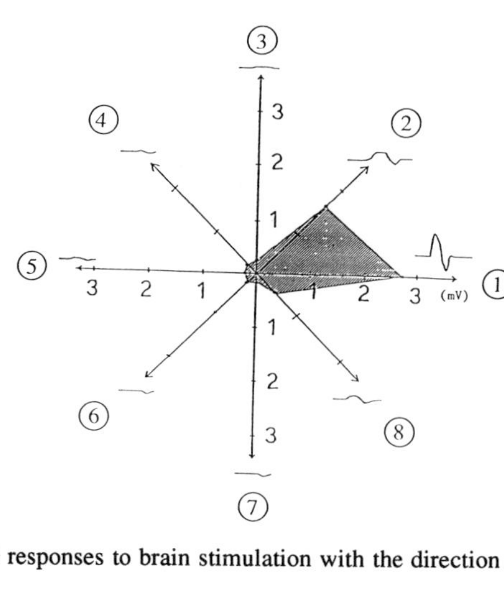

Figure 14. Change of EMG responses to brain stimulation with the direction of stimulating currents. (© 1990 IEEE)

The maximum EMG response was obtained in direction 1, whereas no EMG signals were obtained in directions 3, 4, 5, 6 and 7. Magnetic stimulation of the region of the human cortex related to the foot area was also achieved.

The functional mapping of the motor cortex related to both hand and foot area was obtained as shown in Figure 15. Magnetic brain stimulation was carried out at all grid points of the meshed region on the surface of the head. The distance between the neighboring grid points is 5 mm. The arrows show current directions for neural excitation. An optimal direction of stimulating currents for neural excitation exists in each functional area in the cortex. The stimulating current vectors at the hand and foot areas are oppositely directed.

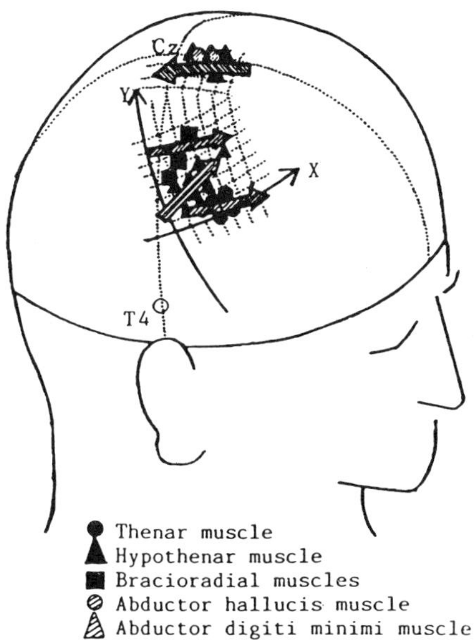

Figure 15. Functional distribution of the human motor cortex related to the hand and foot areas. The arrows show current directions for neural excitation. The distance between grid points is 5 mm. (© 1990 IEEE)

The existence of orientation of current vectors for neural excitation in the brain may be partially understood by the anatomical structures of the neural tissues in the cortex. In other words, it may be possible to estimate both functional and anatomical organization of the neural networks from the mapping of these current vectors for neural excitation. From this point of view, we studied functional mapping of the motor cortex related to the foot area in more detail.

The functional mapping of the motor cortex related to the right and left abductor digiti minimi muscles was obtained as shown in Figure 16. The stimulating currents were directed from the Nasion to the Inion. The contour map of the right hemisphere indicates the functional area that innervates the left abductor digiti minimi muscles, and the contour map of the left hemisphere indicates the functional area that innervates the right abductor digiti minimi muscles. Selective brain stimulation of the right and left foot areas was achieved.

The dashed line in Figure 16 indicates the boundary line that separates the contour maps in the right and left hemisphere. No EMG responses to magnetic brain stimulation were obtained on the dashed line. The dashed line is shifted 3 mm to the right from the Y-axis.

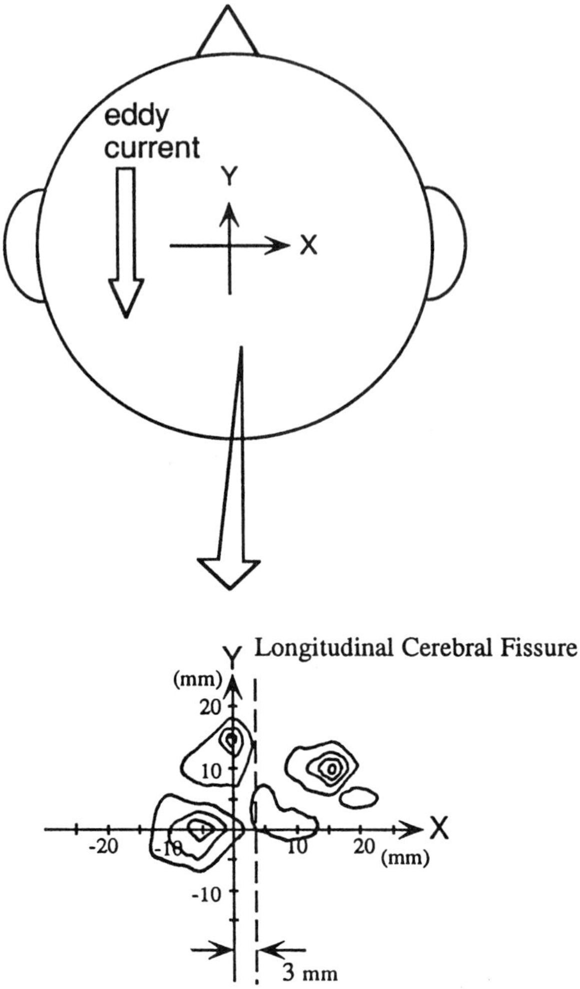

Figure 16. Functional distribution of the motor cortex related to the right and left abductor digiti minimi muscles. The stimulating currents flow in the direction from Nasion to Inion. The isofield contour map was obtained by plotting the equipotential lines of EMG recordings from abductor digiti minimi muscles that responded to magnetic brain stimulation. The curves are quantized by 20 uV from 0 to 80 uV. Y-axis runs on the Nasion-Inion axis, and the zero point is Cz of the international 10-20 system. The dashed line coincides with the longitudinal cerebral fissure.

Figure 17 shows a coronal section of an MRI scan obtained from the same subject as in Figure 16. The MRI data show that the longitudinal cerebral fissure is shifted 3 mm to the right from the Nasion-Inion axis (Y-axis). The MRI in Figure 17 confirms that the dashed line in Figure 16 coincides with the longitudinal cerebral fissure.

We made functional maps of the human brain related to the right and left abductor digiti minimi muscles (ADMs) as shown in Figure 18. Magnetic brain stimulation was carried out in the central areas of both left and right hemispheres. EMG signals from both right and left ADMs were measured simultaneously. When the stimulating currents were oriented in the direction from the right to the left hemisphere, clear EMG signals were obtained only from the right ADM. No EMG responses were obtained from the left ADM, as shown in Figure 18(a).

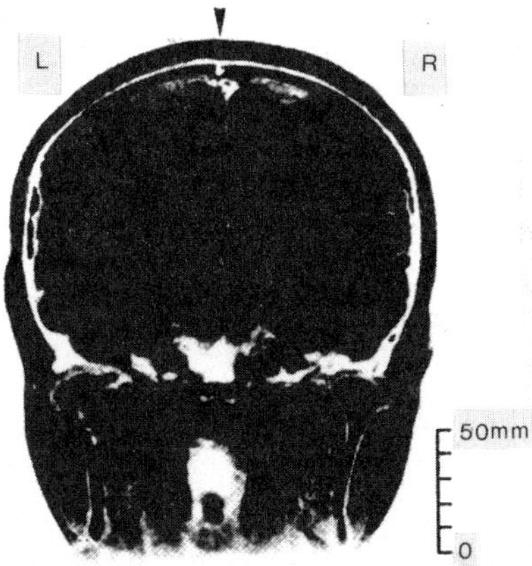

Figure 17. Coronal section of MRI obtained from the same subject in Fig. 16. The MRI shows that the longitudinal cerebral fissure is shifted 3 mm to the right from the marked point at Cz.

The peak points that innervate the right ADM were localized not only in the left hemisphere (Point B, (X,Y) = (-15 mm, 10 mm)), but also in the right hemisphere (Point A, (X,Y) = 15mm, 10mm).

The functional map of the brain related to the left ADM is shown in Figure 18 (b). The stimulating currents were oriented in the direction from the left to the right hemisphere. Clear EMG signals were obtained only from the left ADM, with no EMG responses from the right ADM. The peak points that innervate the left ADM were localized in the right hemisphere: Point C, or (X,Y) = (35mm, 15mm), and Point A, or (X,Y) = (15 mm, 10 mm).

Directional effects of the stimulating currents were studied in detail at points A, B and C, by rotating the figure-of-eight coils and stimulating these points. At Point A, for example, when the stimulating currents flow from the right to the left hemisphere, clear EMG signals were recorded only from the right ADM. These results verify that ipsilateral neural innervation exists in the motor cortex related to the foot region. In contrast, when the stimulating currents were oriented in the direction from the left to the right hemisphere, clear EMG signals were obtained only from the left ADM.

Figure 19 shows the optimal stimulating direction for neural excitation at each point. Broken lines represent the inferred direction of the neural fibers that innervate the right ADM, and solid lines represent the inferred direction of the neural fibers that innervate the left ADM at each point on the surface of the head. These results suggest that the excitation of neural fibers are dependent on the direction of induced eddy currents. As we will discuss later, this vector map may reflect the anatomical direction of the nerve fibers.

MODELING OF NEURAL EXCITATION ELICITED BY MAGNETIC STIMULATION

The existence of orientation in the brain may be partially understood by the anatomical structure of the neural tissues in the cortex. We introduced a model that describes the neural excitation by induced eddy currents.

The model was based on the assumption that a nerve is excited easily when induced eddy currents flow in the direction parallel to the nerve fiber [22]. Reilly also suggested that the component of stimulating currents perpendicular to the nerve fiber has little stimulating effects [27].

Figure 20 shows schematic diagrams which describe the neural excitation by induced eddy currents. When the eddy currents that are induced by time-varying magnetic fields flow in the direction from a soma to the distal parts, as shown in Figure 20(a), depolarized parts are excited, and the membrane excitation propagates along the nerve fiber from the excited parts to the distal parts.

In contrast, as shown in Figure 20(b), when the induced eddy currents flow in the direction from the distal parts to soma, hyperpolarized parts block or inhibit neural excitation even if the depolarized parts near the soma can be excited.

This assumption was partially verified by an experiment of electrical stimulation of the human arm. When the median nerve was stimulated in the direction from the proximal to the distal part, a clear EMG signal was recorded from the thenar muscles. In contrast, when the median nerve was stimulated in the opposite direction, very weak EMG signals were obtained.

We refined the model as shown in Figure 21. When the induced eddy currents flow parallel to the nerve as shown in Figure 21 (a) and 21 (b), their distribution is centered on

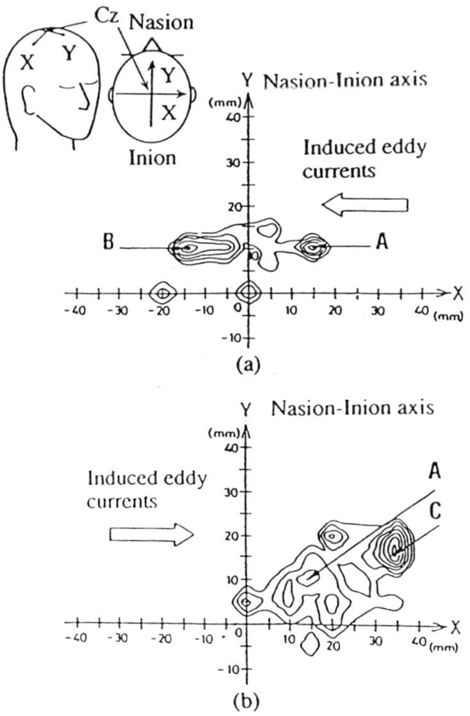

Figure 18. Functional distributions of the motor cortex related to the right abductor digiti minimi muscles (ADM) (a), and the left ADM (b). The Y axis runs parallel to the Nasion-Inion axis, and the zero point is Cz of the international 10-20 system. The isofield contour maps were obtained by plotting the same amplitude of electromyographic (EMG) responses measured from the ADMs. (© 1991 IEEE)

Point C. Point A is depolarized, and Point B is hyperpolarized. The membrane excitation propagates along the nerve fiber from the excited region to the distal region. When the nerve is inclined to the induced eddy currents as shown in Figure 21(c), Point A is excited. The membrane excitation propagates along the nerve fiber from the excited region to the distal region with ease. In contrast, when the nerve is inclined to the induced eddy currents as shown in Figure 21(d), induced eddy currents cause Point B to become hyperpolarized. The potential of Point A is too low to cause nerve excitation, the nerve being harder to excite than in the case shown in Figure 21 (b) and (c).

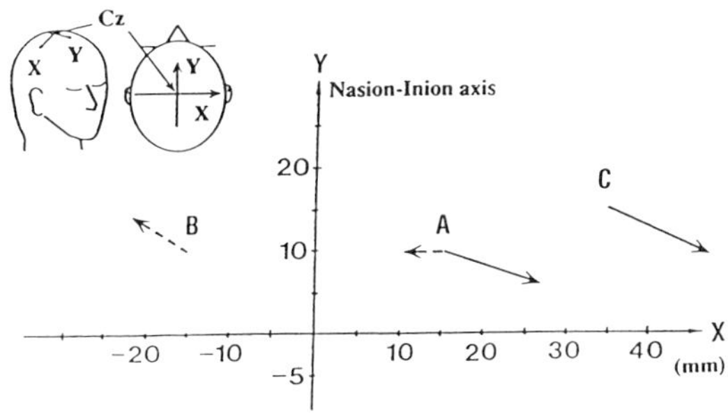

Figure 19. Optimal directions for neural excitation at Point A, B and C in Figure 18. The solid-line arrows represent the inferred directions of neural fibers that innervate the left ADM, and the dotted-line arrows represent the inferred directions of neural fibers that innervate the right ADM. (© 1991 IEEE)

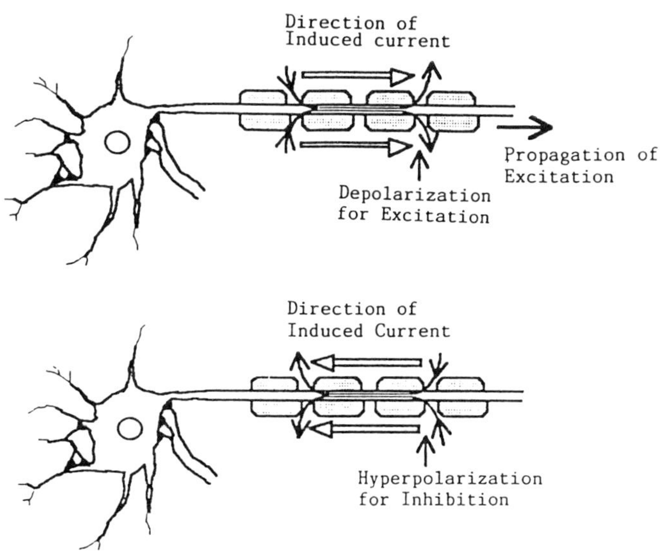

Figure 20. Direction of stimulating current and neural excitation. (a) The induced eddy currents flow in the direction from soma to distal parts. Depolarized parts are excited, and the membrane's excitation propagates along the myelinated nerve axon from the excited parts to the distal parts. (b) The induced eddy currents flow in the direction from distal parts to soma. Hyperpolarized parts inhibit neural excitation. (© 1990 IEEE)

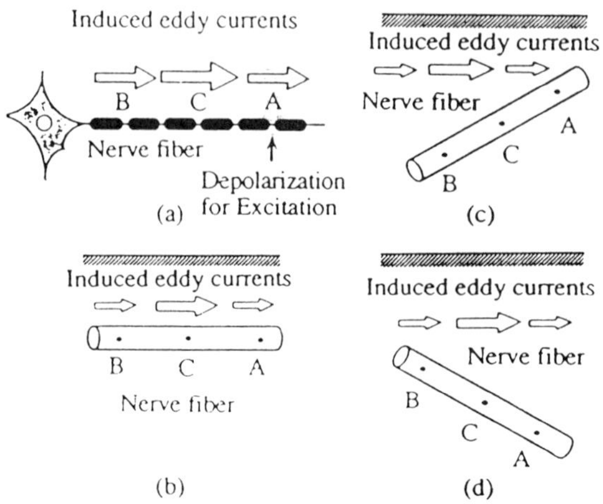

Figure 21. Direction of induced eddy current and neural excitation. When a nerve fiber is oriented parallel to the surface of the body as shown in (a) (b), Point A is excited. When a nerve fiber is inclined as shown in (c), Point A is easily excited. When a nerve fiber is inclined in the opposite direction as shown in (d), the nerve is not so easily excited. (© 1991 IEEE)

We simulated the excitation of neural fibers as a function of direction of angle between figure-of-eight coil and nerve axon using the parameters of the Frankenhaeuser-Huxley model [28]. We applied Roth and Basser's model [29] to our method of neural excitation elicited by focal and vectorial stimulation with a figure-of-eight coil. We considered a figure-of-eight coil with 30 turns having a radius of 25 mm at each coil. The distance between the center of the coil and the nerve is 27.6 mm. We represented the coil as a 64-sided polygon, solved the integral analytically, and then added the contributions to obtain the induced electric fields.

The change of neural membrane potential elicited by magnetically induced eddy currents is shown in Figure 22. The duration of stimulating current was 100 μs. The X' axis lies in a direction parallel to the neural fiber, and the zero point lies beneath the intersection of the figure-of-eight coil. Stimulating currents were oriented parallel to the X axis.

When the angle between the figure-of-eight coil and the nerve is 45 degrees, as in Figure 22(a), the nerve becomes depolarized at point X' = (20 mm), and the nerve is hyperpolarized at point X' = (-10 mm). If the hyperpolarized area does not draw down the depolarization, the nerve can be excited at X' = (20 mm). The computer simulation shows that the membrane potential of the hyperpolarized area returns rapidly to the resting potential and does not affect neural excitation at X' = (20 mm). The excitation propagates both sides of the neural fiber.

In contrast, when the angle between the figure-of-eight coil and the nerve is -45 degrees, the area of hyperpolarization is more intense at X' = (-20 mm), as shown in Figure 22(b), and the membrane potential remains below the threshold of excitation. These simulated results suggest that the direction of neural excitation is governed by the angle of inclination of the neural fibers themselves.

Although the simple model for a single neuron fails to recognize the convoluted structure of the cerebral cortex, the modified model is valid for the explanation of the effect of the vectorial characteristics of induced eddy currents on neural excitation. In other words, the model simulates the change of neural excitation processes with the change of angle between the figure-of-eight coil and the nerve fiber.

The validity of the model was partly verified by experiments using a frog's isolated nerve bundle stimulated by magnetically induced electric fields. A sciatic nerve bundle dissected from a frog was positioned in a saline solution. Magnetic nerve stimulation was

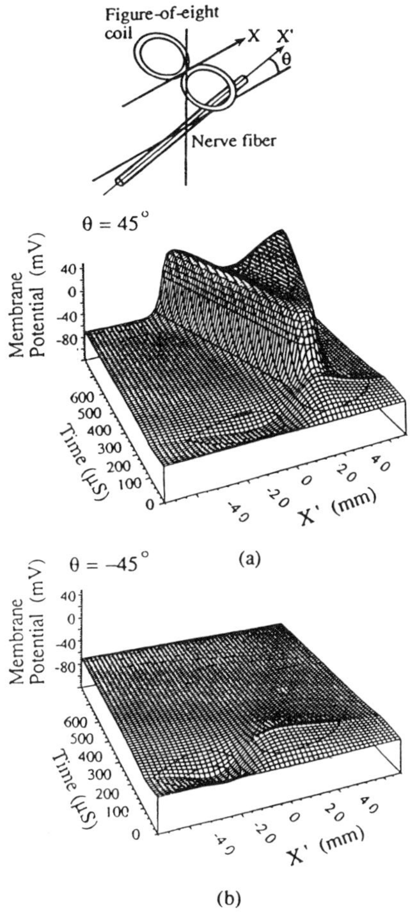

Figure 22. Simulation of nerve excitation elicited by magnetically induced eddy currents. The X axis runs parallel to the direction of the tangent of both circular coils. In this simulation, currents in the coil flow along the -X axis at the intersection. The X' axis runs in a direction parallel to the nerve fiber. When the angle between the figure-of-eight coil and nerve is 45 degrees, as shown in (a), the nerve is excited at point X' = (20 mm), and the excitation propagates to both sides of the nerve fiber. In contrast, when the angle between the figure-of-eight coil and the nerve is -45 degrees, as shown in (b), the nerve fiber is not excited.

carried out, changing the angle between the figure-of-eight coil and the nerve bundle. When the angle was 45 degrees, the threshold for neural excitation was 9.5 V/m. In contrast, when the angle was -45 degrees, the threshold was 15 V/m.

The existence of the vectorial characteristics of stimulating currents for neural excitation reflects both the functional and anatomical organization of neural fibers in the brain.

ACKNOWLEDGEMENTS

The author wishes to thank Dr. T. Matsuda and Dr. O. Hiwaki, Department of Electronics, Kyushu University, and Dr. M. Fujiki, Department of Neurosurgery, Medical College of Oita, for their cooperative work.

The work was supported in part by grants from the Ministry of Education, Science and Culture in Japan.

REFERENCES

1. A. D'Arsonval, Dispotifs pour la mesure des courants alternatifs des toutes frequences, *C.R. Acad. Sci.*, 48: 450-451 (1896).
2. R.G. Bickford and B.D. Fremming, Neuronal stimulation by pulsed magnetic fields in animals and man, *Dig. 6th Int. Conf. Med. Electr. Biol. Eng.* 112 (1965).
3. D.D. Irwin, S. Rush, R. Evering, E. Lepeschkin, D.B. Montgomery, and R.J. Weggel, Stimulation of cardiac muscle by a time-varying magnetic field, *IEEE Trans. Magnetics* MAG-6, 321-322 (June 1970).
4. J.A. Maass and M.M. Asa, Contactless nerve stimulation and signal detection by inductive transducer, *IEEE Trans. Magnetics* MAG-6, 322-326 (June 1970).
5. P.Å. Öberg, Magnetic stimulation of nerve tissue, *Med. Biol. Eng.* 11: 55-64 (1973).
6. S. Ueno, S. Matsumoto, K. Harada and Y. Oomura, Capacitative stimulatory effect in magnetic stimulation of nerve tissue, *IEEE Trans. Magnetics* MAG-14, 958-960 (Sept. 1978).
7. S. Ueno, P. Lövsund and P.Å. Öberg, On the effect of alternating magnetic fields on action potential in lobster giant axon, *Proceedings of the 5th Nordic Meeting on Med. and Biol. Eng., Linköping, Sweden,* 262-264 (1981).
8. S. Ueno, P. Lövsund and P.Å. Öberg, Effects of time-varying magnetic fields on the action potential in lobster giant axon, *Med. & Biol. Eng. & Comput.* 24: 521-526 (1986).
9. S. Ueno, K. Harada, C. Ji, and Y. Oomura, Magnetic nerve stimulation without interlinkage between nerve and magnetic flux, IEEE Trans. Magnetics, MAG-20, 1660-1662 (Sept. 1984).
10. A.T. Barker, R. Jalinous, and I.L. Freeston, Non-invasive magnetic stimulation of human motor cortex, *The Lancet,* i.: 1106-1107 (May 1985).
11. A.T. Barker, I.L. Freeston, R. Jalinous, and J.A. Jarratt, Clinical evaluation of conduction time measurements in central motor pathways using magnetic stimulation of the human brain, *The Lancet,* i.: 1325-1326 (June 1986).
12. A.T. Barker, I.L. Freeston, R. Jalinous, and J.A. Jarratt, Magnetic stimulation of the human brain and peripheral nervous system: An introduction and results of an initial clinical evaluation, *Neurosurgery,* 20 (No. 1): 100-109 (Jan. 1987).
13. B.L. Day, P.D. Thompson, J.P. Dick, K. Nakashima, and C.D. Marsdes, Different sites of action of electrical and magnetic stimulation of the human brain, *Neuroscience Letters,* 75: 101-106 (1987).
14. C.W. Hess, K.R. Mills, and N.M.F. Murray, Measurement of central motor conduction in multiple sclerosis by magnetic brain stimulation, *The Lancet,* 397-419 (1987).
15. C.W. Hess, K.R. Mills, and N.M.F. Murray, Responses in small hand muscles from magnetic stimulation of the human brain, *J. Physiol.*, 388: 397-419 (1987).
16. K.R. Mills, N.M.F. Murray, and C.W. Hess, Magnetic and electrical transcranial brain stimulation, Physiological mechanisms and clinical applications, *Neurosurgery,* 20 (No.1): 164-168 (Jan. 1987).
17. J.C. Rothwell, B.L. Day, P.D. Thompson, J.P.R. Dick, and C.D. Marsden, Some experiences of techniques for stimulation of the human cerebral motor cortex through the scalp, *Neurosurgery,* 20 (No. 1): 156-163 (Jan. 1987).
18. S. Ueno, T. Tashiro and K. Harada, Localized stimulation of neural tissues in the brain by means of paired configuration of time-varying magnetic fields, *J. Appl. Phys.* 64: 5862-5864 (Nov. 1988).
19. S. Ueno, T. Matsuda, M. Fujiki and S. Hori, Localized stimulation of the human motor cortex by means of a pair of opposing magnetic fields, *Digests of Intermag Conf.*, GD-10, Washington D.C., (March 1989).
20. S. Ueno, T. Matsuda and M. Fujiki, Localized stimulation of the human cortex by opposing magnetic fields, *in:* "Advances in Biomagnetism," S.J. Williamson et al., eds., Plenum Press, New York and London (1989).
21. S. Ueno, T. Matsuda and M. Fujiki, "Localized stimulation of the human brain by a pair of opposing pulsed magnetic fields, *Memoirs of the Faculty of Engineering - Kyushu University,* 49: 161-173 (Sept. 1989).
22. S. Ueno, T. Matsuda and M. Fujiki, Functional mapping of the human motor cortex obtained by focal and vectorial magnetic stimulation of the brain, *IEEE Trans. on Magn.,* 26 (No. 5): 1539-1544 (Sept. 1990).
23. S. Ueno, T. Matsuda and O. Hiwaki, Localized stimulation of the human brain and spinal cord by a pair of opposing pulsed magnetic fields, *J. Appl. Phys.* 66: 5838-5840 (1990).
24. S. Ueno, T. Matsuda and O. Hiwaki, Estimation of structures of neural fibers in the human brain by vectorial magnetic stimulation, *IEEE Trans. Magn.* 27 (No. 6): 5387-5389 (Nov. 1991).
25. S. Ueno, T. Tashiro, S. Kamise, T. Oosako and K. Harada, A paired-coil configuration for localized hyperthermia of deep tissues, *Digests of Intermag Conf.*, BD-3, St. Paul, Minnesota, (April 1985).
26. S. Ueno, T. Tashiro, S. Kamise and K. Harada, Localized hyperthermia by means of a paired-coil configuration: calculation of current distributions in cubical model, *IEEE Trans. Magn.*, MAG-23: 2437-2439 (Sept. 1987).
27. J.P. Reilly, Peripheral nerve stimulation by induced electric currents: exposure to time-varying magnetic fields, *Medical & Biol. Eng. & Comput.* 27: 101-110 (March 1989).

28. B. Frankenhaeuser and A.L. Huxley, The action potential in the myelinated nerve fiber of Xenopus Laevis as computed on the basis of voltage clamp data, *J. Physiol.* 171: 302-315 (1964).
29. B.J. Roth and P.J. Basser, A model of stimulation of a nerve fiber by electromagnetic induction, *IEEE Trans. on Biomed. Eng.* 37 (No. 6): 588-597 (June 1990).

DISCUSSION

Chairman Prof. M. Markov: Thank you, Professor Ueno, for an excellent presentation. Are there some questions and comments? Please.

Dr. J.A. Nyenhuis: Dr. Ueno, can you say how critical the intensity of the stimulating pulse is when you map out the motor cortex?

Prof. Ueno: Do you mean the critical threshold for nerve excitation?

Dr. Nyenhuis: That's right.

Prof. Ueno: We didn't measure the induced eddy current density directly, but compared to electrical stimulation, the threshold for nerve excitation is about $1mA/cm^2$ to $5mA/cm^2$.

Dr. Nyenhuis: So if you increase the intensity of the stimulator by say, 20 percent, do you still get the same detailed mapping of the brain?

Prof. Ueno: Now I understand your important question. The answer is partially yes, and partially no. In mapping out the motor cortex, the direction of stimulating currents, or the direction of eddy currents, is very critical. The brain maps vary with the direction of stimulating current. This is very critical. But if you fix the direction of the stimulating current, you obtain almost the same brain maps even if you change the intensity of the stimulator by 10 to 20 percent.

Prof. Markov: Other questions? Let me make a comment. I appreciate what you presented regarding the influence of eddy currents, but it seems to me very unlikely that this is the only possible explanation of the influence of magnetic fields. It seems to me that the direct influence of the magnetic field on the axon or other structures in the brain should also be considered.

Prof. Ueno: That is a most basic question. We have been working with the pure magnetic field effect on nerve excitation for more than 10 years, but we have not yet obtained positive results. I would like to mention Dr. Schwartz's work. Before our lobster experiment, Dr. Schwartz also used lobster giant axons to study possible effects of static magnetic fields on nerve excitation processes. In his experiment, instead of the intracellular recording by microelectrodes, he introduced a sucrose method for measuring action potentials extracellularly. He measured action potentials during DC or static magnetic field exposures, up to 1.2 tesla. He obtained no positive effects of the magnetic field.

Therefore, it may be concluded that DC fields up to 1.2 tesla or 2 tesla do not affect nerve excitation processes.

Prof. Markov: Any other questions or comments? If not, then thank you again, Professor Ueno.

MOTOR CORTICAL STIMULATION IN MAN

J.C. Rothwell

MRC Human Movement and Balance Unit
Institute of Neurology
Queen Square
London WC1N
United Kingdom

PHYSIOLOGICAL PRINCIPLES OF MAGNETIC STIMULATION

Magnetic stimulation can be regarded as a convenient way of getting electric current to flow within the brain. As such, it might be expected to be very similar to conventional methods of brain stimulation using external electric stimuli. However, from a physiological standpoint, there is an important difference between the two techniques. Magnetic and electric stimulation induce different directions of current flow within the brain, and this can have an important influence on which neuronal elements are activated preferentially.

Neurones in the brain are arranged in a very specific pattern, and are oriented in particular directions relative to the cortical surface. The major output neurones of any area of the cortex are the pyramidal neurones. These lie perpendicular to the surface of the brain and have long axons that run into the white matter below the cortex. Many interneurones, in contrast, are oriented parallel to the surface of the brain and have relatively short axons that run within the cortical grey matter.

A neurone is activated when a stimulus produces electric current that flows outwards across its membrane from the inside to the outside. For a neurone with a long axonal process, this current is proportional to the the spatial derivative of the electrical field along the length of the axon. Clearly, an electrical stimulus from a focal anode will be capable of stimulating virtually any element in the cortex, since the current flows in most directions away from the anode, both radially and tangentially to the surface of the skull. Magnetic stimulation, on the other hand, produces electrical currents that flow only parallel to the surface of the skull. Therefore radially oriented neurones are unlikely to be stimulated by this method. Activation will favour neurones with axons that tend to run parallel to the surface of the brain.

Unfortunately, there is one more complexity to add to this picture. The surface of the brain is not flat. It is folded considerably. Thus, although the neurones maintain their orientation with respect to the surface of the brain, this varies with respect to the surface of the skull. The net result is that because of this folding, it is almost impossible to predict which type of neurone will be stimulated best by electrical or magnetic stimulation. Nevertheless, the argument is important, because it is an experimental fact that different types of neurones do indeed seem to be activated preferentially by the two forms of stimulation.

This can be demonstrated very effectively after stimulation of the motor cortex. It is easy to obtain EMG responses in contralateral body muscles after electrical or magnetic stimulation of the motor cortex. However, the responses to the two forms of stimulation differ in one important way. At low to medium intensities of stimulation, electrical pulses produce muscle activation with a latency about 1 or 2 ms shorter than the responses seen after magnetic stimulation. In physiological terms, that is quite a significant length of time, as 1 or 2 ms is time enough for a nerve action potential to travel 12 cm or more along the length of axon. Why should magnetic and electric stimulation of the brain activate the motor cortex in different ways?

There are two possible explanations for the difference. First, it may be that electric stimulation activates the pyramidal neurones directly, producing what is known as a D-wave of descending activity in the corticospinal tract (Patton & Amassian, 1954.) On a simple model of the motor cortex, the pyramidal neurones are oriented perpendicular to the surface and are well stimulated by radial current flow. On this model, as shown in Figure 1, magnetic stimulation will fail to activate these cells, but might instead activate other neurones that run parallel to the surface of the brain and excite pyramidal neurones trans-synaptically. The latency difference in the muscle responses that is seen in the periphery would simply be due to the synaptic transmission delay within the motor cortex (Day et al, 1989; Amassian et al, 1990.)

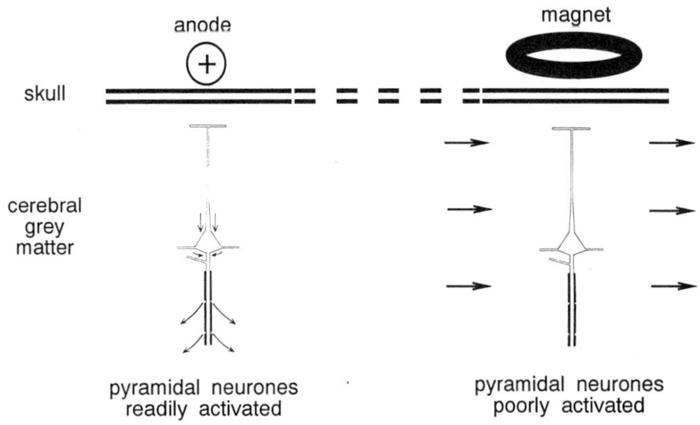

Figure 1. Relationship between the orientation of pyramidal neurones in the cortical grey matter and the direction of current flow induced in the brain by anodal electrical or magnetic stimulation. On the left, anodal currents are drawn entering the cell body of the neurone and leaving at nodes along the length of the axon. On the right, the current induced by magnetic stimulation is shown flowing parallel to the surface of the skull at right angles to the pyramidal neurone. This direction of current flow is ineffective in activating the cell.

This simple model can be criticised because the orientation of the motor cortex varies with respect to the surface of the skull due to the folding of the brain. The second explanation is that electrical stimulation activates the axons of pyramidal cells very deep down in the brain, some 12 cm or so from the surface. In contrast, magnetic stimulation may activate the pyramidal cells within the cortex itself. The latency difference between the two forms of stimulation would be due to the conduction time for magnetically evoked impulses to travel down the first 12 cm of the axon (Edgley et al, 1990.)

Whichever explanation for the latency difference is correct, both have a very important consequence that we can make use of in physiological experiments. Electric stimulation activates pyramidal cell axons in the subcortical white matter. At this point, their excitability is uninfluenced by the level of excitation in the motor cortex itself. Thus, whether the cortex is excited or inhibited, responses to electric stimulation should always be the same. With magnetic stimulation, excitation occurs within the grey matter of the cortex. Here, the excitability of the pyramidal cells depends on the level of cortical excitability, being easier to activate if the cortex is excitable, and more difficult to activate if the cortex is inhibited. Thus, the responses to magnetic stimulation reflect very much the excitability of the cortex itself.

There is one proviso to this argument. The electric stimulus must be small if it is to activate the axons of pyramidal tract cells preferentially. If the stimulus is large, then cortical interneurones are also activated. Practically, this means that such experiments are performed during a background voluntary activation of the muscle under test. In this state, the threshold for cortical stimulation is lowest, and hence the intensity of the electrical stimulus can be at a minimum.

When a magnetic stimulator is used to activate the motor cortex to produce a response in a muscle, the stimulus first excites corticospinal fibres, which then conduct impulses to the spinal cord. At this point, there is a synapse onto the spinal motoneurones, which innervate the muscle from which we are making the EMG record. Thus, any change in the excitability of either the spinal cord or the cortex may change the size of the muscle response that we see. If a small electric stimulus is used to activate the cortex, then the muscle response should only change size because of changes in the level of spinal cord excitability. By combining both techniques in man, we can obtain information about the level of motor cortex excitability.

INPUTS TO THE MOTOR CORTEX FROM OTHER AREAS OF THE BRAIN

The ability to analyse the excitability of the motor cortex itself allows us to investigate some of the inputs to that area of the brain from other regions of the central nervous system. Experiments have been described showing inputs from muscle stretch receptors in the periphery to the motor cortex (Day et al, 1991), and from other areas of the brain, including the cerebellum (Ugawa et al, 1991), the contralateral motor cortex (Ferbert et al, 1992), and neighbouring areas of the ipsilateral motor cortex itself (Kujirai et al, 1992.) In this report, I shall concentrate on the latter two effects as examples of this technique.

Figure 2 shows the technique for investigating transcallosal connections between the two sides of the brain. In this experiment, two magnetic stimulating coils are used. One, the conditioning coil, is used to activate, say, the right motor cortex. The other stimulator is used to test the excitability of the left motor cortex at different times after the conditioning shock. It turns out that a stimulus to the right side of the brain can inhibit responses evoked from stimulating the left motor cortex at intervals of 6 to 7 ms or longer.

The EMG traces from the right first dorsal interosseous muscle in the top of Figure 3 illustrate the effect in the raw data from one subject. The first trace shows a control response to the test stimulus (of the left motor cortex) given on its own. In the lower traces, the test stimulus has been preceded by a conditioning stimulus given at different times beforehand to the opposite side of the head. When the conditioning stimulus was given 6 or 7 ms before the test stimulus, the size of the test response was inhibited. There is quite a large change, and it is not necessary to average many trials to see the effect.

Figure 2. Schematic diagram of the arrangement of magnetic stimulating coil used for demonstrating transcallosal inhibition between the motor cortices. Two relatively focal figure-of-eight coils (Ueno et al,1988) are used over the lateral part of each hemisphere. The figure is drawn as if looking down on the head from above (nose forward).

The lower graph is the time course of inhibition averaged from seven subjects. In this graph, the size of the control response is set to be 100 percent. Values less than this indicate that the test response was inhibited by the conditioning shock. The X-axis indicates the interval between the conditioning and the test stimuli. Inhibition begins at about 5 or 6 ms, and then it begins to recover by 15 ms or so. The duration of inhibition depends upon the intensity of the conditioning stimulus. The higher the conditioning intensity, the longer the period of inhibition.

As stated above, it is important to know at what level in the central nervous system (spinal cord or cortex) this inhibition is occurring. For example, the conditioning stimulus could be inhibiting spinal cord mechanisms so that a descending test volley produced a smaller EMG response in the muscle because the spinal cord was less excitable than usual. On the other hand, the conditioning stimulus may have inhibited the cerebral cortex itself, so that a given test stimulus evoked a smaller descending volley from cortex to cord than in the control situation. We think that the latter is true, because the same conditioning stimulus has no effect on test responses produced by an anodal electric stimulus.

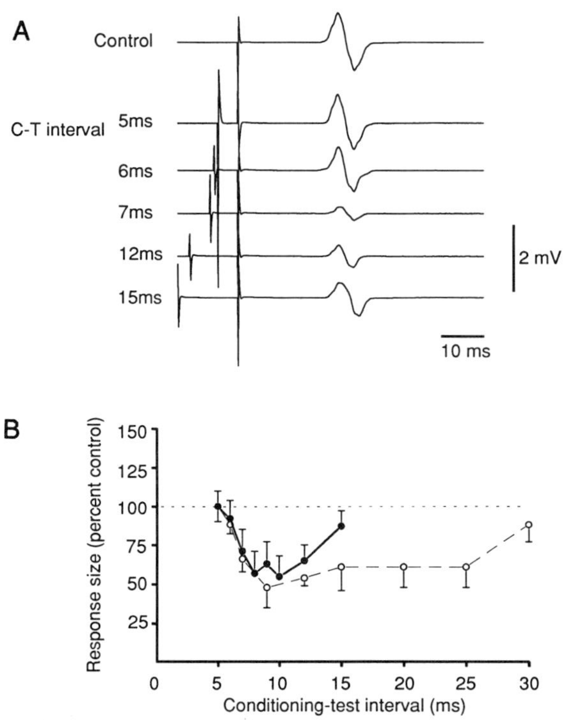

Figure 3. A : Effect of a conditioning magnetic stimulus delivered by a figure-of-eight coil (peak magnetic field rating of 2.4T) over the left motor cortex on EMG responses in the left first dorsal interosseous muscle produced by a magnetic test shock delivered by a second identical coil over the right hemisphere. The top trace shows the response to the test shock given alone. The lower five traces illustrate the effect of a conditioning shock given 5, 6, 7, 12 or 15 ms before the test shock (C-T interval). The traces are aligned to the onset of the test shock. Each trace is the average of 10 single trials. The subject was relaxed throughout. The intensity of both the test and the conditioning stimulus was 55 percent of maximum output.

B : Mean ±1SEM time course of interhemispheric inhibition in six different normal subjects using two different intensities of conditioning shock. Responses were recorded in the relaxed FDI muscle, and the size of the conditioned response at each interval has been expressed as a percentage of the size of the test shock given alone (=100%). The conditioning shock was set to be about 10 (•) or 25 (o) percent above the threshold for producing responses in contralateral relaxed hand muscles. As in the subject in A, both test and conditioning stimuli were delivered through figure-of-eight coils. (From Ferbert et al, 1992, with permission.)

This is illustrated in Figure 4. These experiments were carried out in active muscles, so that the intensity of the test shocks could be as small as possible. The upper EMG traces show examples of the effects on one subject. The lower graph shows the average effects on four different subjects. Responses to electric stimulation of the brain were not inhibited by a conditioning stimulus to the opposite hemisphere; responses to magnetic stimulation were suppressed considerably. From this result, we conclude that the conditioning stimulus was having an inhibitory effect on the excitability of the contralateral motor cortex.

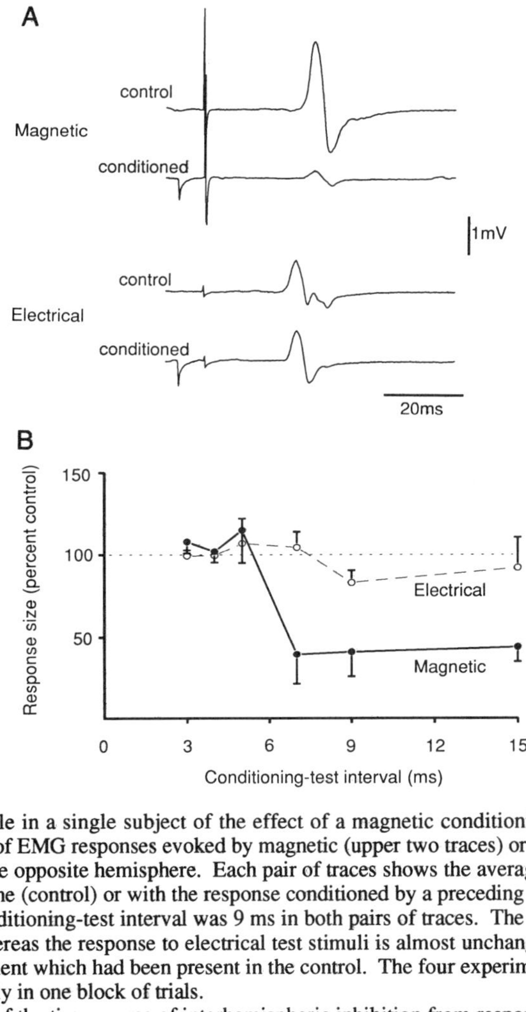

Figure 4. A : Example in a single subject of the effect of a magnetic conditioning stimulus over one hemisphere on the size of EMG responses evoked by magnetic (upper two traces) or electrical (lower two traces) stimulation of the opposite hemisphere. Each pair of traces shows the average response (of 10 trials) to the test shock given alone (control) or with the response conditioned by a preceding magnetic shock (conditioned). The conditioning-test interval was 9 ms in both pairs of traces. The response to magnetic test stimuli is inhibited, whereas the response to electrical test stimuli is almost unchanged, apart from loss of a second positive component which had been present in the control. The four experimental conditions were applied pseudo-randomly in one block of trials.

B : Comparison of the time course of interhemispheric inhibition from responses elicited by electrical or magnetic test stimuli in four subjects. Points are mean ±SEM (paired t-test of comparing mean inhibition of electrically and magnetically evoked test responses showed significant ($P<0.05$) differences at intervals of 7, 9 and 15 ms). Magnetic conditioning stimuli were given via a figure-of-eight coil having a maximum output rating of 2.4T. Stimulation intensity ranged from 50-67 percent. Magnetic test stimuli were given through a large round coil having a maximum output rating of 2T. Stimulation intensities ranged from 45-74 percent. The mean peak-to-peak size of the control EMG response evoked by magnetic stimuli (2.7mV) was slightly larger than control responses evoked by electrical test stimuli (2.2mV). (From Ferbert et al, 1992, with permission.)

The final piece of the jigsaw is the pathway by which this inhibition is mediated. That is, how did the effect cross from one hemisphere to the other? Because of the time interval at which inhibition becomes noticeable, it is likely that the pathway involves the corpus callosum. There are very many fibres that link the two hemispheres of the brain across the corpus callosum, and we think that the inhibitory effect seen in these experiments was conducted via this pathway. Indeed, we have examined one patient with agenesis of the corpus callosum, and in her the inhibition was absent (Rothwell et al, 1991).

Having demonstrated the effect, and the pathway by which it is mediated, two questions remain about transcallosal inhibition. First, what is its normal function, and second, how does this go wrong in neurological disease? As an answer to the first question, it is possible to investigate how transmission in this transcallosal pathway changes during the performance of different types of movement. Preliminary experiments show that the amount of transcallosal inhibition is greater if the subject is performing a movement with the hand contralateral to the conditioning hemisphere, whilst keeping the other hand relaxed, as compared to a task in which the subject uses both hands at the same time. In other words, it appears that the brain controls bimanual hand movements differently than unilateral hand movements. It may be that the transcallosal inhibitory pathway is useful in preventing movement of one hand when we use the other.

This transcallosal pathway also appears to go wrong in neurological disease. Patients with myoclonus have muscle jerks that may occur in response to a sensory stimulus, or may occur when a patient tries to move. In some cases the jerks affect the whole body, so that, for example, a tap to the arm on one side will cause a massive jerk of both the arms and legs of the patient. In some of these patients, it appears that the somatosensory stimulus provokes an abnormal discharge on the contralateral side of the brain, which then spreads via the corpus callosum to drive activity in the other hemisphere (Brown et al, 1991). The net result is a bilateral twitch of the body. In this sort of patient, transcallosal inhibition seems to be absent. Indeed, in some cases the normal inhibition is replaced by a facilitation, consistent with the known pathophysiology of these patients.

Another example of using two magnetic stimulators to explore connections between different parts of the brain is the demonstration of possible cortico-cortical inhibitory connections between different parts of the motor cortex in the same hemisphere. The experiment is a variation on the transcallosal test described above. However, in this case, the conditioning coil is placed at the vertex of the skull in order to stimulate the leg area of the motor cortex.

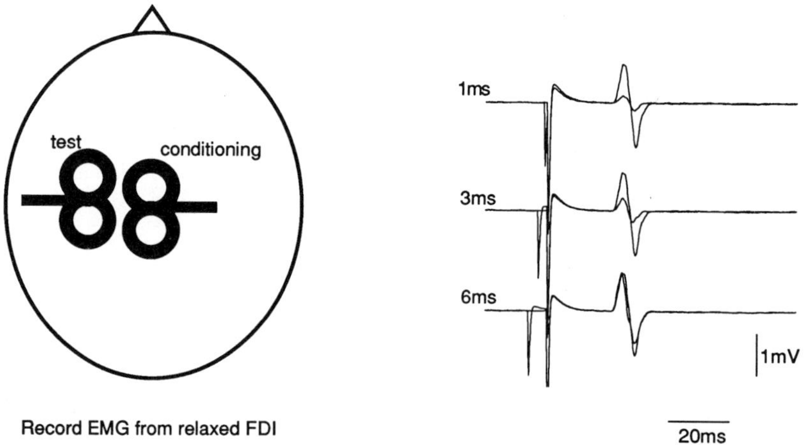

Figure 5. Probable cortico-cortical inhibition from a conditioning stimulus placed over the vertex of the skull to a test coil placed over the hand area of the motor cortex. The right hand traces show averaged EMG records from the relaxed first dorsal interosseous muscle of the left hand of a typical normal subject. Each trace consists of two superimposed records: the response to the test stimulus given alone, and the response conditioned by the vertex stimulus (numbers on the left of each trace.) The control response is the larger of each pair.

The test coil is again on the hand area of the motor cortex. Figure 5 illustrates the experimental arrangement and some typical EMG records from a normal subject.

In these experiments, the intensity of the test shock is set to be such that it evokes responses of about 1 to 2 mV peak-to-peak amplitude in contralateral relaxed hand muscles. The intensity of the conditioning shock is set so that it will elicit responses in leg muscles, but is incapable of producing any responses in hand muscles, even if the subject activates the hand. Under these circumstances, if the conditioning stimulus is given 1 ms to 6 ms before the test stimulus, then the test response is inhibited. If the conditioning stimulus is moved away from the vertex, then this inhibition disappears.

We can demonstrate that the inhibition is probably occurring at a cortical level if we compare the effect of the vertex conditioning shock to responses evoked by anodal electric test stimuli to the lateral part of the motor cortex. As shown in Figure 6, responses to anodal test stimuli are unaffected by conditioning stimuli applied to the vertex, whilst the responses to magnetic stimulation are suppressed.

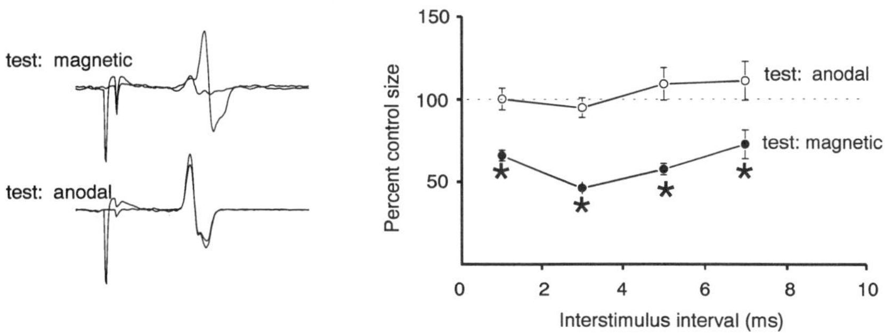

Figure 6. Comparison of the effect of a vertex conditioning shock on responses evoked by a magnetic or anodal electric test shock in the active first dorsal interosseous muscle of one subject. Traces on the left are raw data showing the effect at a conditioning-test interval of 2 ms. Each trace consists of two superimposed records: the average response (of 10) to the test stimulus given alone and the response when conditioned by the vertex shock. In each trace, the control response is the larger of the two. Note that the response to the magnetic test shock is almost completely suppressed by the conditioning stimulus, whilst the response to anodal stimulation is virtually unaffected. The graph on the right shows the time course of this effect. Each point is the mean (±1SE) of 10 observations. Asterisks indicate a significant ($P<0.05$) difference in the effect of the conditioning stimulus or the two types of test response.

This type of inhibition is clearly quite different from the transcallosal inhibition described previously. It begins at very much shorter interstimulus intervals, and is elicited from a different position of the conditioning coil. We can think of it as a type of "surround inhibition" of one area of cortex by another. Again the question is, does this circuitry ever go wrong? We have not yet had an opportunity to investigate this question in detail. However, one natural condition to examine is epilepsy, in which patients are thought to have abnormalities of both excitatory and inhibitory mechanisms within the brain. Recent data collected in this laboratory suggest that abnormalities of the inhibitory circuit may exist in certain groups of epileptic patients.

In conclusion, magnetic stimulation is a useful method of activating the brain of normal subjects through the intact skull. However, the current flow induced in the brain differs from that produced by a conventional electric stimulator by being limited to directions parallel to the surface of the skull. Because neurones have particular orientations in the brain, and because they are differentially sensitive to the direction of the stimulating current, then magnetic and electric stimuli can at times activate different structures. In the motor cortex it seems that the responses of muscles to magnetic stimulation are much more influenced by the level of excitability within the cortex than the responses produced by low intensity electric stimulation of the brain. This difference can be very useful when we wish to examine the effect of inputs that we suspect might modulate cortical excitability in man.

ACKNOWLEDGEMENTS

The work described in this article is the result of a collaboration with very many colleagues over several years: Drs. Brian Day, Philip Thompson, Kaz Ugawa, Andreas Ferbert, Alberto Priori, Jim Colebatch, Peter Brown, Takashi Kujirai, Alain Maertens de Noordhout, and Professors David Marsden and Pat Merton. Mr. Richard Bedlington provided the firm bedrock of background support.

REFERENCES

1. V.E Amassian, G.J. Quirk and M. Stewart, A comparison of corticospinal activation by magnetic coil and electrical stimulation of monkey motor cortex, *Electroencephalography and Clinical Neurophysiology* 77: 390-401 (1990).
2. P. Brown, B.L. Day, J.C. Rothwell, P.D. Thompson and C.D. Marsden, Intrahemispheric and interhemispheric spread of cerebral cortical myoclonic activity and its relevance to epilepsy, *Brain* 114: 2333-2352 (1991).
3. B.L. Day, D. Dressler, A. Maertens de Noordhout, C.D. Marsden, K. Nakashima, J.C. Rothwell and P.D. Thompson, Electric and magnetic stimulation of human motor cortex: Surface EMG and single motor unit responses, *Journal of Physiology* 412: 449-473 (1989).
4. B.L. Day, H. Riescher, A. Struppler, J.C. Rothwell and C.D. Marsden, Changes in the response to magnetic and electrical stimulation of the motor cortex following muscle stretch in man, *Journal of Physiology* 433: 41-57 (1991).
5. S.A. Edgley, J.A. Eyre, R.N. Lemon and S. Miller, Excitation of the corticospinal tract by electromagnetic and electrical stimulation of the scalp in the macaque monkey, *Journal of Physiology* 425: 301-320 (1990).
6. A. Ferbert, A. Priori, J.C. Rothwell, B.L. Day, J.G. Colebatch and C.D. Marsden, Interhemispheric inhibition of the human motor cortex, *Journal of Physiology* 453: 525-546 (1992).
7. T. Kujirai, J.C. Rothwell, J. Fong, P.D. Thompson and B.L. Day, Local intracortical inhibition in human sensorimotor cortex, *Journal of Physiology* (proceedings, in press) (1992).
8. H.D. Patton and V.E. Amassian, Single and multiple unit analysis of cortical stage of pyramidal tract activation, *Journal of Neurolophysiology* 17: 345-363 (1954).
9. J. C. Rothwell, J.G. Colebatch, T.C. Britton, A. Priori, P.D. Thompson, B.L. Day and C.D. Marsden, Physiological studies in a patient with mirror movements and agenesis of the corpus callosum, *Journal of Physiology* 438: 34P (1991).
10. S. Ueno, T. Tashiro and K. Harada, Localized stimulation of neural tissues in the brain by means of a paired configuration of time-varying magnetic fields, *Journal of Applied Physics* 64: 5862-5864 (1988).
11. Y. Ugawa, B.L. Day, J.C. Rothwell, P.D. Thompson, P.A. Merton and C.D. Marsden, Modulation of motor cortical excitability by electrical stimulation of the cerebellum in man, *Journal of Physiology* 441: 57-72 (1991).

DISCUSSION

Chairman Prof. M. Markov: For this exciting paper, now it is time for questions and comments. Please.

Dr. M. Fujiki: Three years ago, we reported similar data on conditioning the system. If my understanding is correct, in the bi-phasic facilitating phase, the first facilitation is the H reflex and the second is the so-called long latency reflex. Please let me hear your opinion of this.

Dr. Rothwell: I think you are absolutely right and, as you say, a peripheral input first of all excites the spinal cord so the mono-synaptic reflexes are increased, and then it excites the cortex, so that the cortical responses are increased as well. Yes, I agree entirely with you, of course. There are effects at two levels with this system. But it's very nice that with these techniques, you can sort them out. You can use H reflexes to look at the spinal cord and magnetic stimulation will help you look at the cortex.

Prof. S. Matsuoka: Could you tell me why you use two machines for magnetic stimulation? Usually electrical double stimulation is only from the one stimulator. Is there a problem?

Dr. Rothwell: Yes, the problem is that these machines will only give stimuli once every 3 seconds. So, if you want to give stimuli that are only 3 milliseconds apart, you have to have two machines. And I think it's also a more difficult engineering problem to get one of these machines to give trains of very rapid stimuli, because the power requirements are enormous.

Prof. Matsuoka: Could you tell me, are there any variances in classical localization in a motor cortex followed by magnetic stimulation? I mean in epilepsy patients.

Dr. Rothwell: No, motor cortex localization looks normal in the patients that we studied.

Prof. Matsuoka: I mean that usually I have two cases for the multiple sub-dural implanted electrode. Vertically there is no problem but horizontally there are many variances.

Dr. Rothwell: I haven't got a lot of experience with this technique.

Prof. Matsuoka: Thank you.

Prof. Y. Mano: Thank you very much, very nice presentation. I am Mano from Nara Medical University, Department of Neurology. We do similar experiments in double magnetic stimulation, especially stimulation to the cerebellum and cortical stimulation. And it's as you said, there is inhibition in the motor-evoked potential. But, do you think there are several pathways between the cerebellum and the cortex and also to the descending tract?

Dr. Rothwell: As you saw, I left out cerebellar stimulation because of those particular reasons. You can stimulate over the cerebellum and get inhibition, but to prove that the inhibition actually is produced by stimulation of the cerebellum and not some other structure is very difficult.

Prof. Mano: The strength of the stimulation is probably one clue, but in your presentation you didn't say much about strength of stimulation in the cortex. Even if you produce a very minor stimulus to the cerebellum you get several inhibitions.

Dr. Rothwell: Yes, I agree entirely, but like you said, it's difficult to know what the pathway is.

Prof. Mano: Well, one point is stimulation with time difference, and one is stimulation strengths, especially the conditioning strengths, so it might be clarified.

Prof. Markov: Other questions? Comments? Dr. Rothwell, according to your experience, at this time of day, what do we need more: magnetic stimulation or food stimulation?

Dr. Rothwell: Food, please.

Prof. Markov: I would like to thank all the speakers for excellent papers, and to thank all of you for attending, and we now close this section of the symposium for lunch break.

CLINICAL APPLICATION OF TRANSCRANIAL MAGNETIC STIMULATION FOR INTRAOPERATIVE MONITORING OF THE SPINAL CORD AND MAPPING OF THE MOTOR CORTEX

Karl H. Kraus[1], Walter J. Levy[2], Lavern D. Gugino[1], Rhamsis Ghaly[3], Vahe Amassian[4], John Cadwell[5]

[1]Brigham and Womens Hospital, Harvard School of Medicine
75 Francis Road, Boston, Massachussetts, U.S.A.

[2]Cleveland Clinic of Florida,
3000 Cypress Road, Ft. Lauderdale, Florida, U.S.A.

[3]Dept. of Neurological Surgery, University of Illinois
Chicago, Illinois, U.S.A.

[4]Dept. of Physiology, State University of New York
Brooklyn, New York, U.S.A.

[5]Cadwell Industries, Kennewick, Washington, U.S.A.

INTRODUCTION

Since the first descriptions of the use of time-varying magnetic fields for non-invasive stimulation of nervous structures, this modality has been suggested as a versatile tool for studying and monitoring the nervous system [1, 2]. Much has been recently learned about the mechanisms of magnetic stimulation, its safety, and its potential applications. Studies in normal volunteers and those with neurologic disorders have led to important findings of the organization of the motor system in health and disease [3-13].

This paper addresses two areas of magnetic stimulation that are needed because of their great utility, and will offer new understandings of brain injury and organization. These uses have opposite engineering requirements. Operative monitoring requires large motor cortical areas to be stimulated for stable potential recording from several muscles concurrently, and cortical mapping requires the smallest possible area for discrete stimulation of specific cortical areas. We will consider each.

INTRAOPERATIVE SPINAL CORD MONITORING

Monitoring of the spinal cord has become standard of care for those surgeries that place the spinal cord at risk. To date, the most common monitoring modalities are somatosensory evoked potentials and spinal evoked potentials. Somatosensory evoked potentials are carried in the dorsal columns of the spinal cord which carry sensory information. Since motor function preservation is often of paramount importance, this modality can only be an indirect indi-

Biomagnetic Stimulation, Edited by
S. Ueno, Plenum Press, New York, 1994

cator of motor function. Additionally, the vascular supply to the posterior columns is separate from the anterior columns and gray matter. Therefore, somatosensory evoked potential monitoring also indirectly monitors vascular supply to the spinal cord motor system. False positive and false negative results from somatosensory evoked potentials have been frequently reported [14-18].

Spinal evoked potentials have also been used to monitor the spinal cord. Since the spinal cord is grossly stimulated, there is some question as to which tracts, motor, sensory, or both, are truly being stimulated. If the evoked potential is recorded caudal to the stimulating site, there is the possibility of antidromal conduction in sensory fibers. Additionally, spinal evoked potentials in most situations cannot be performed in awake patients, excluding the opportunity for awake base line recording and monitoring during patient positioning. Also, most somatosensory and spinal evoked potential monitoring configurations only record potentials that follow white-matter pathways, while the gray matter of the spinal cord is most susceptible to ischemia and injury.

Transcranial magnetic motor evoked potential monitoring has the advantages that it can be performed on awake individuals for preoperative assessment and during patient positioning, and monitors motor system pathways through the anterior gray horn of the spinal cord. It can be combined with somatosensory evoked potential monitoring for a more complete assessment of spinal cord function.

Transcranial magnetic motor evoked potential monitoring has a disadvantage in that stable signals are difficult to achieve in operating room situations. One specific reason has been in coil design. Using conventional round coil designs, it is often difficult to stimulate bilateral upper and lower limb motor areas simultaneously. This is because the magnetic coil stimulates most intensely under the coil windings, and currently available round coils cannot physically cover the entire motor area.

For this reason, a cap-shaped coil was developed, which can overlie a majority of the motor cortex bilaterally. Since this coil is in the shape of a cap, it is called the "cap coil."

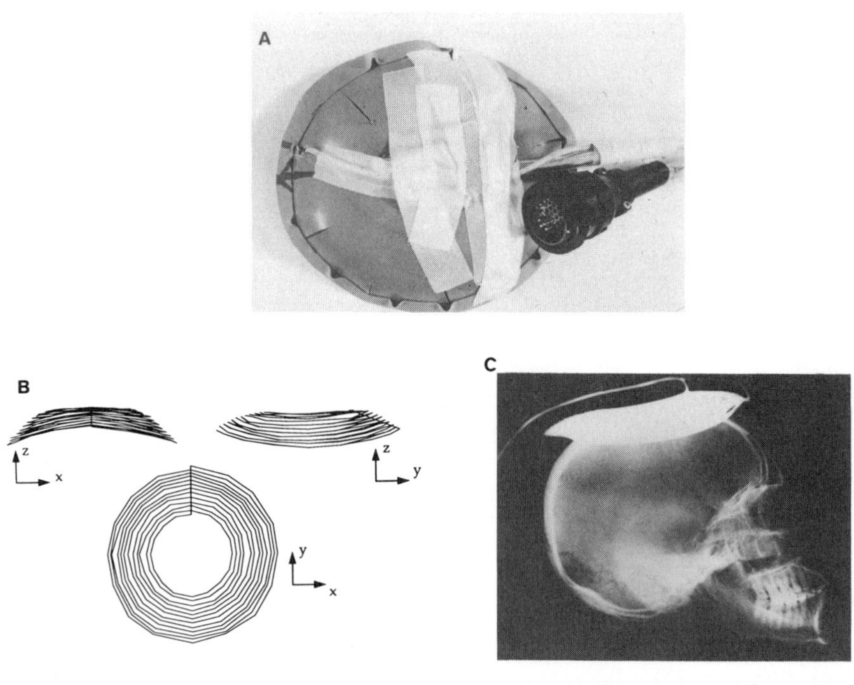

Figure 1. Three visual perspectives of a cap-shaped coil. Part A is a photograph of the coil. Part B is a computer-generated image of the coil windings. Part C is a photograph of a radiograph of the coil placed on a skull.

Figures 1A-C give a visual perspective of the cap coil. Figure 1A is a photograph of the prototype of the cap coil. Figure 1B is a digitized computer figure of the coil windings. The cap coil consists of 12 turns of copper wire encased in a rubber compound for insulation. The outer diameter of the coil is approximately 17 cm, with the central diameter of 7 cm devoid of wire. Figure 1C is a photograph of a radiograph of the cap coil placed on a skull.

The proper position of the cap coil should be so that the coil windings overlay the motor cortex. It is important to understand that the maximum stimulation of a magnetic coil occurs under the coil windings, not the geometric center of the coil. In order to demonstrate proper position of the cap coil, motor cortical mapping was performed in an individual and depicted in Figure 2. This technique involves focally stimulating the motor cortex with a figure-of-eight-shaped coil by placing this coil at specified locations on the scalp. The technique and application of motor cortical mapping will be more fully described in the second part of this paper.

The outlined areas of the figurine are scalp areas which, when stimulated with a figure-of-eight coil, resulted in compound muscle action potentials in the left and right abductor digiti minimi (ADM) and anterior tibialis (ATM) muscles. The wave forms on either side of the figurine are the resulting compound muscle action potentials following 80 percent full stimulus intensity stimulation using a Cadwell MES-10 and the cap coil.

The skyscraper graph at the top of Figure 2 gives the amplitudes of the resulting compound muscle action potentials. The coordinate 0,0 relates to Cz of the 10-20 system of EEG electrode placement.

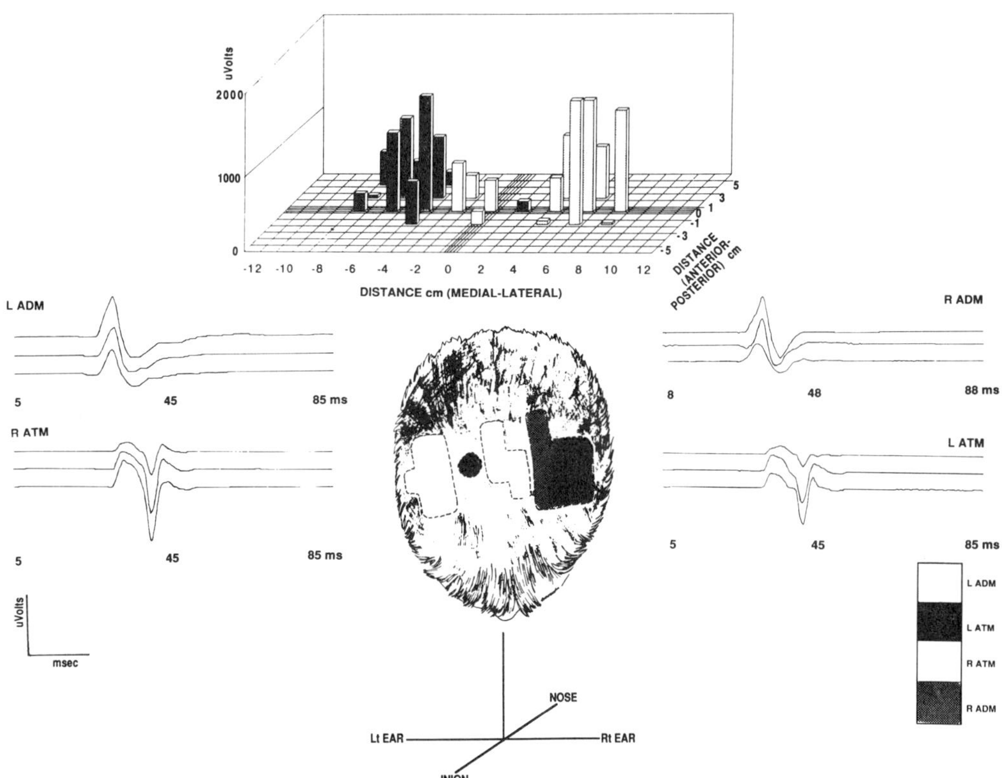

Figure 2. Location of scalp areas which, when stimulated with focal magnetic stimulation, result in compound muscle action potentials of the left and right abductor digiti minimi (ADM), and anterior tibialis (ATM) muscles. The shaded areas of the figurine are the scalp locations that correspond to the legend of muscles at the lower right corner of the figure. To the left and right of the figurine are compound muscle action potentials recorded from these muscles. The skyscraper graph above the figurine reflects the areas as well as the amplitudes of the compound muscle action potentials.

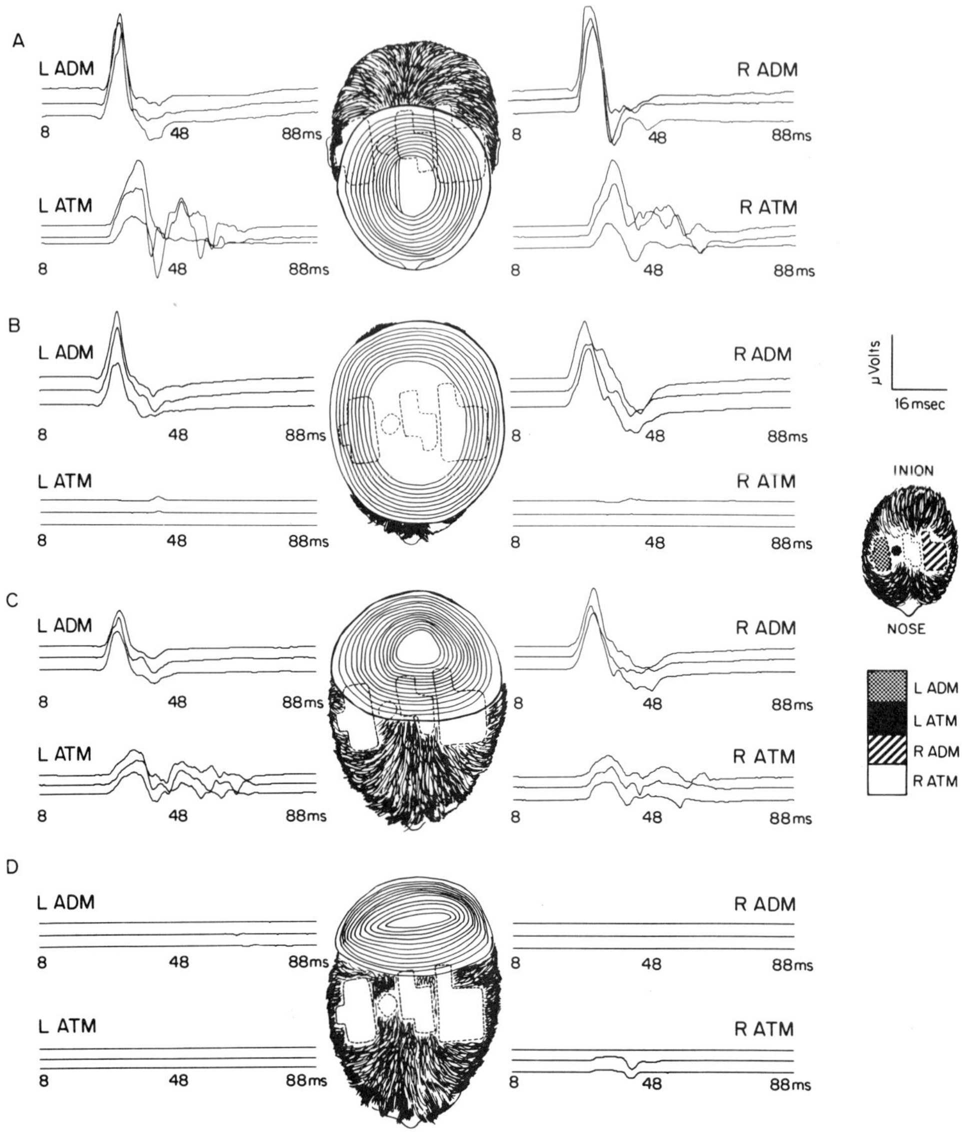

Figure 3. The result of positioning the cap coil (depicted by the coil's winding) in four positions related to the cortical map of Figure 2. To the left and right of the figurine are resulting compound muscle action potentials from three cap coil stimulations at that position.

Figure 3A-D demonstrates compound muscle action potential amplitude for the left and right ADM and ATM as a function of cap locations for the same volunteer whose data was depicted in Figure 2. In the center of the figure, four cap positions on the scalp are depicted: A) anterior with cap rim 2 cm above the nasion; B) centered on vertex; C) anterior rim at the vertex; and D) cap anterior rim 3 cm posterior to vertex. The cap locations are depicted as coils of wire superimposed over the scalp areas found to elicit compound muscle action potentials for the left and right ADM and ATM depicted in Figure 2.

The gray scale shown at the right of the figure is used for identifying the scalp muscle focus defined by the figure-of-eight coil. To the left and right of each scalp figurine are examples of compound muscle action potentials retrieved from the four muscles labeled when stimulated at 80 percent full power of the MES-10. Vertical calibration bar is equal to 1,000 μ volts. Negativity is up for all traces. Sweep is shown in msec shown for each set of three stacked responses. In each set of responses, sweep begins at 8 msec for artifact elimination.

Position A, where the cap is in an anterior position, and where the coil windings cover the motor cortical areas identified by focal stimulation techniques, is the most beneficial for soliciting compound muscle action potentials from all four muscles.

Monitoring motor pathways to several limbs is advantageous if the procedure may affect one, but not all, of the pathways monitored. If evoked potentials are lost in the pathway coursing through the operative site, and not those outside the operative site, then anesthetic, temperature, drug and systemic blood pressure effects may be excluded. The coil can be fixed to the patient's head with an elastic band so that it remains in the same position during monitoring. Movement of a magnetic coil often results in changes in monitored compound muscle action potentials. Fixing the coil to the head also allows more access to the patient's head area by the anesthesia personnel.

In Figure 4, the calculated induced electrical field of the cap coil is shown. The graph is the magnitude of the induced electrical field across a plane 1 cm below the bottom edge of the coil as a function of position.

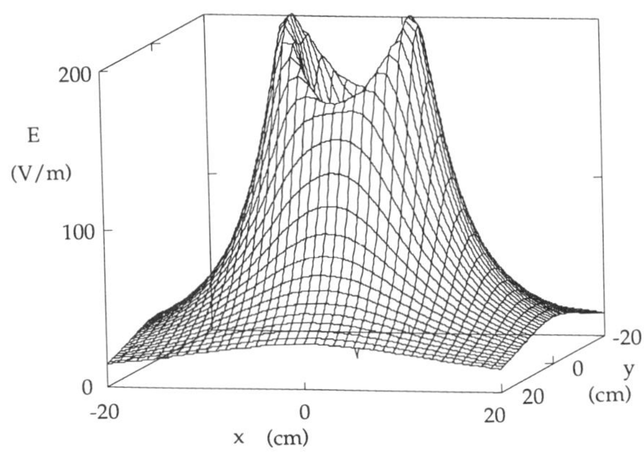

Figure 4. The cap coil's induced electric field strengths calculated on a plane 1 cm below the coil. Notice that the volcano appearance demonstrates a largely symmetric field whose tip is in the center of the band of windings placed over the scalp. There is a fore and aft peak in the volcano which is about 15 to 20 percent above the remainder of the peak at the top of the activation area.

Figure 5 compares the induced electrical field as a function of depth for the cap coil, a 9 cm round coil, and a figure-of-eight coil. The intensity of the magnetic stimulation of each of the coils was adjusted to that the induced electrical field was equivalent between them at the scalp surface. As is shown, the cap coil is not only capable of covering more of the motor cortex, but also induces large electrical fields. Additionally, the cap coil is able to induce greater electrical fields at all depths. Larger induced electrical fields tend to result in more consistent evoked potentials.

A factor determining the success of any intraoperative motor system monitor is use of an anesthetic technique that is permissive for eliciting recoverable and consistent evoked potentials. A number of investigators have reported that transcranial magnetic stimulation induced compound muscle action potentials are difficult to obtain during general anesthesia. All inhalation anesthetics used currently have been found to markedly attenuate transcranial magnetic compound muscle action potentials. These included Ethrane, Isoflourane, Halothane, and Nitrous Oxide in concentrations greater than 50 percent by volume [19-24].

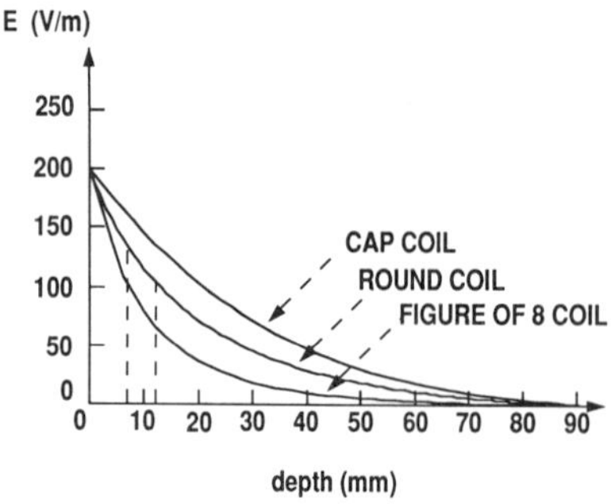

Figure 5. Induced electrical fields were calculated for three different coil designs: the cap coil, a 9-cm round coil, and a figure-of-eight coil. Calculated electrical fields were adjusted so that they were equivalent directly beneath the coils. Vertical broken lines suggest the approximate location of the surface of the motor cortex.

Nitrous oxide in concentrations less than 50 percent has been shown to be compatible with transcranial magnetic induced compound muscle action potentials useful for monitoring in the monkey and man [20, 25].

Studies have also been reported on the effects of intravenous agents on transcranial magnetic stimulation induced compound muscle action potentials. The potent narcotics sufentanyl and fentanyl, which are used in many anesthetic techniques, are permissive. Hypnotic anesthetic agents, such as the benzodiazepines, barbiturates, and propofol, have been shown to severely depress induction of compound muscle action potentials [26, 27]. Recent reports have found that the neuroleptanesthetic techniques, etomidate and ketamine, when serving as the primary anesthetic technique, are compatible with retrieving transcranial magnetic stimulation induced compound muscle action potentials [27-30]. Based on a comparison of effects, ketamine and etomidate were both superior to the neuroleptanesthetic techniques. A note of interest is that both of these agents have been shown to enhance central nervous system excitability.

A second concern with respect to choosing a permissive anesthetic technique involves the ubiquitous use of muscle relaxants. Vecuronium, for example, will not change the latency of transcranial magnetic induced compound muscle action potentials up to an 80 percent block, but amplitudes will be affected with much lesser myoneural blocks [31]. Compound muscle action potential amplitudes will remain unchanged up to a 40 percent block, but will decrease by 50 percent at an 80 percent block. Since both compound muscle action potential amplitude and latency are typical measures followed during central nervous system monitoring, continually assessing the degree of peripheral muscle relaxant should be stressed [25]. Quantitative methods for monitoring degree of myoneural blockade are available.

Based on previous studies, the anesthetic technique employed by the authors consisted of 1) etomidate infusions serving as the primary hypnotic, 2) Fentanyl infusions for analgesia, 3) nitrous oxide no greater than 50 percent by volume in oxygen, and 4) an infusion of atracurium for muscle relaxation. The degree of peripheral myoneural junction relaxation was continually monitored in order to maintain a block which varied between 30 and 40 percent.

In the first 25 patients monitored using this anesthetic technique, none showed movements during the maintenance phase of anesthesia, and none reported an incidence of intraoperative recall of surgical events. Recently, an alternate protocol has been used which consists of 1) propofol and ketamine infusions serving as hypnotics, 2) Fentanyl infusions for analgesia, and 3) infusions of atracurium for muscle relaxation. These two protocols allow adequate amplitude in transcranial induced muscle action potentials for accurate monitoring.

Following is a case which reflects the utility of concurrent somatosensory and motor evoked potential monitoring. An 18-year-old man was operated on for coarctation of the aorta which had been unsuccessfully operated on 12 years previously. An angiographic examination had demonstrated two areas of constriction in the thoracic descending aorta. A Doppler echocardiography demonstrated a pressure gradient of 40 mmHg across the first constriction.

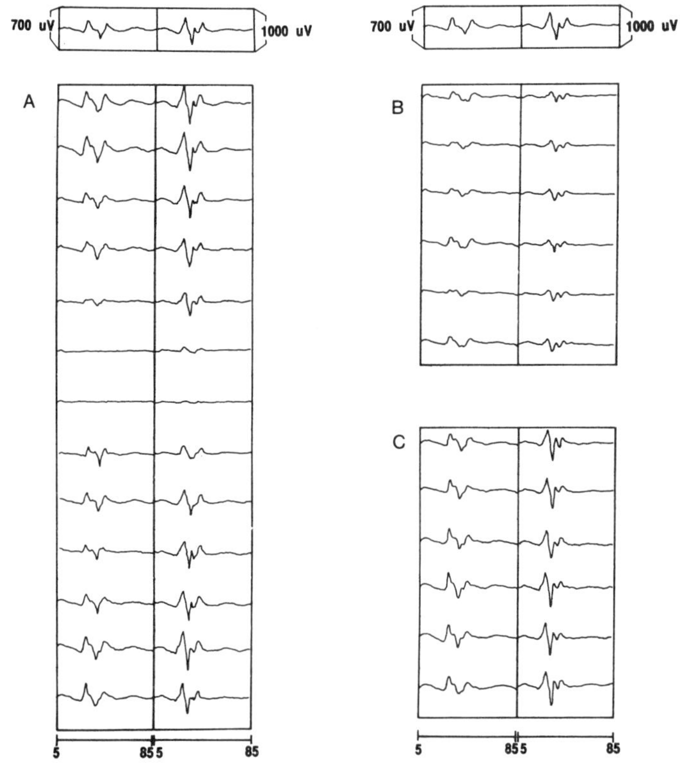

Figure 6. Monitoring of a patient with aortic coarctation with transcranial motor-evoked potentials. Depicted are compound muscle action potentials from the left (left column) and right (right column) anterior tibialis muscles. The evoked potentials were collected before, during and after a five-minute complete occlusion of the descending aorta (part A). Three minutes after cross clamping, motor-evoked potentials disappeared. The surgeon removed the clamp, and within 10 minutes the MEPs returned. The SEP (not shown) showed no change. The surgeon placed a bypass shunt, which reduced blood flow to the descending aorta, and resumed the aortic graft procedure (part B). Following coarctation repair, evoked potentials returned to original amplitudes (part C) and the patient recovered with no neurologic deficits.

The patient was premedicated with fentanyl and induced with sufentanil (10 μg/kg); etomidate (0.5 mg/kg) and atracurium for intubation. Anesthetic maintenance included infusions of etomidate (0.5 mg/kg/hr), sufentanil (2 μg/kg/hr), and atracurium (0.3 to 0.5 mg/kg/hr) with oxygen. A Datex Relaxograph (Datex Instrumentation Corporation, Helsinki, Finland) was used for monitoring the degree of atracurium induced muscle relaxation which was maintained at a 40 to 50 percent block level.

Spinal cord monitoring consisted of both somatosensory and motor evoked potentials. SEPs were monitored with stimulation of both left and right tibial nerves at the ankle. Square wave pulses of 100 μsec at 75 mA with a repetition rate of 3.71 Hz were used. The evoked potentials were recovered with scalp electrodes placed at Cz and Fz for both limbs. Analog filters were set with a lower limit of 10 Hz and an upper limit of 300 Hz. One hundred repetitive stimulations constituted a single average. Optimal digital bandwidths of 10 to 110 Hz were used for both posterior tibial nerve elicited cortical responses.

Motor evoked potentials were elicited by transcranial magnetic stimulation. A Cadwell MES-10 magnetic stimulator (Cadwell Laboratories, Kennewick, Washington) was used as the power source and the cap coil was used as the stimulator. The cap coil was fixed to the patient's head so that the posterior aspect of the coil was positioned over the pre-central

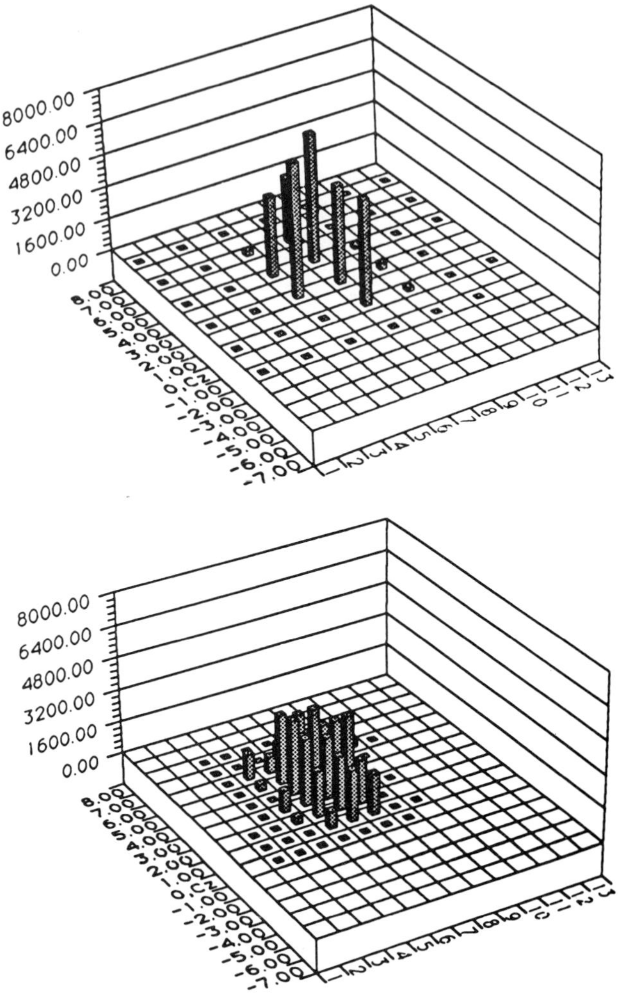

Figure 7. The responses of abductor digiti minimi EMG from a transcranial stimulation map are seen, and the amplitude of the EMG is plotted at each scalp point where stimulation for that muscle occurred. Higher amplitudes of responses indicate stronger muscle activation. It is of importance that these two maps, taken six months apart in the same subject, show areas of stimulation which are within 2 cm^2 of each other. There are scaling differences in the plots which produce different apparent heights of the skyscraper plots, but the EMG amplitudes are similar.

gyrus. The intensity of the magnetic stimulator was set at 80 percent of the maximal intensity allowed by the power source. The active electrodes were placed on the motor point of the anterior tibialis muscles with the reference placed over the proximal insertion tendons. Analog filters for the recovered wave forms were set at a lower limit of 10 Hz and an upper limit of 10,000 Hz.

Bilateral muscle compound action potentials were elicited by single transcranial magnetic stimuli. An automated evoked potential monitoring program was used in this case. The automated program only allows four recording channels to be used; therefore, median nerve SEPs were not included. The monitoring program was allowed to rotate through the three stimulating areas every two minutes. After a stable plane of anesthesia was reached and before surgical intervention, 10 consecutive averages were gathered. Single responses for motor evoked potentials were averaged and means and standard deviations for the principal components of each of the four recording channels were calculated. These served as baselines for subsequent recordings. Automatic alarms were set to occur when the collected evoked potential amplitudes or latencies deviated more than four standard deviations from baseline means.

The evoked potential amplitudes and latency changes obtained from the left and right anterior tibialis compound muscle action potentials are depicted in Figure 6. After exposure of the coarctation sites at 3 hours 20 minutes into the surgery, a complete 5 minute test occlusion of the descending aorta was performed. After 3 minutes, motor evoked potentials were reduced to 50 µV (less than 1 percent of original amplitudes) on the right and could not be recovered from the left anterior tibial muscle. The right MEP wave form latencies changed by 2 µsec (98 percent of original latency.) Neither latency nor amplitudes of somatosensory evoked potentials changed. The test clamps were then removed and the MEPs returned to 50 percent of the original amplitudes after 10 minutes. Based on the MEP amplitude reduction during the test clamp period, the surgeons elected to repair the coarctation with placement of a polytetraflourethylene bypass across the constricted areas.

At 3 hours and 45 minutes into the surgery, tangential aortic occlusion clamps were placed on the aorta to allow for the placement of the implant and to allow partial aortic blood flow. At this time the mean femoral artery pressure dropped to 50 mmHg. During this time the amplitude of the motor evoked potentials remained lower than preocclusion values, though MEP latency, and SEP wave forms remained unchanged. After the partial occlusion clamps were released and until the end of the surgery, the MEP amplitudes steadily increased. The patient recovered uneventfully and without neurologic deficits.

In this case, a 5-minute complete aortic occlusion resulted in a rapid decrease in amplitude of the MEPs, but not the SEPs. The rapid response of MEPs to aortic occlusion has been noted by others, and is consistent with the increased sensitivity of spinal cord gray matter to ischemia.

CORTICAL MAPPING

Cortical mapping is an important goal for transcranial stimulation. Preoperative planning and a wide variety of functional mapping tests could occur with its development. There are other preoperative tests, such as MRI or the PET scan, and the latter shows functional information on a metabolic level. But the completion of our assessment armamentarium requires a test of the electrical circuitry itself. This has historically only been available when the cranium was removed during surgery.

The identification of the location of functional cortex before surgery with respect to a brain tumor or other lesion can determine the approach of a surgeon in order to avoid functional cortical tissue and gain access to the lesion. We might assume that the traditional rules of thumb regarding cortical maps can be used, such as identification of the central sulcus. However this is not always easy to visually identify. It can be obscured or even obliterated by edema around a lesion, or by the lesion itself.

If one assumes there is no distortion induced by the lesion, there is still a variation in the central sulcus, with anatomical variations in which it is small or interrupted in its topography. The most important limitation to recognize is that the motor cortex is not guaranteed to be on the precentral gyrus when it can be identified. There is a variation in the location of the motor homunculus. This was noted even in old studies, but it is observed as well in contemporary mapping studies. This means that the only sure identification of critical motor control areas comes from the stimulation of that brain location and the recording of motor movement in the supplied limb. The somatosensory evoked potential is often used. Here the locating of the area of phase reversal should demarcate the division between sensory and motor cortex, but even this electrophysiological sensory indicator is limited by the appearance of some motor units in overlap with the sensory system.

How can we identify motor cortex with transcranial stimulation? There are three keys. A relatively focal stimulation, a method of correspondence between the scalp site stimulated and the brain, and finally a knowledge of the clinical and functional consequences of the stimulation. The focal stimulation method is available, in the form of the butterfly or double loop coil, first described by Ueno in 1988, and produced by Cadwell Laboratories and several other groups [32]. This coil produces a more focused field than the round coils because the

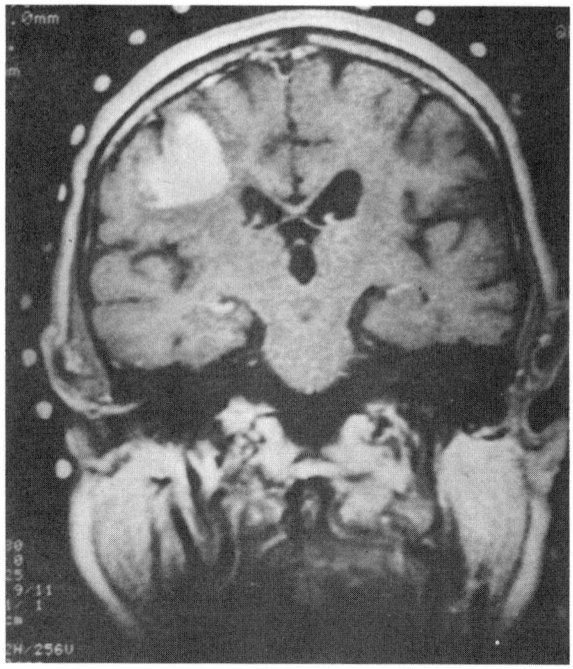

Figure 8. A tumor is seen in this MRI scan, with a grid of capsules across the head that serve as reference landmarks on the MRI scan. Measurements for the MRI scan and for the cortical mapping are made from these same common reference points. The tumor is seen underlying the motor cortex.

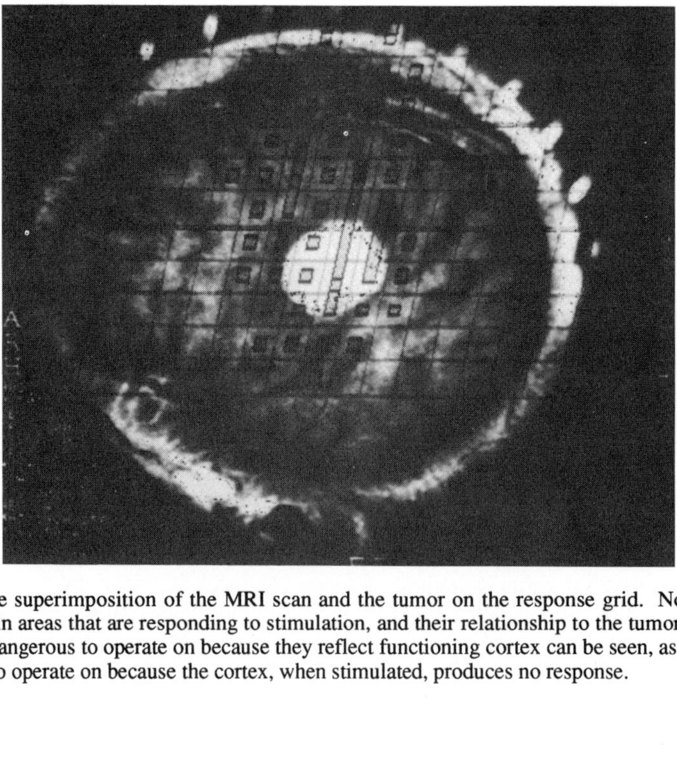

Figure 9. The superimposition of the MRI scan and the tumor on the response grid. Now the maps show those brain areas that are responding to stimulation, and their relationship to the tumor. The locations that would be dangerous to operate on because they reflect functioning cortex can be seen, as can those that would be safe to operate on because the cortex, when stimulated, produces no response.

field at the conjunction of the coils exceeds that at the perimeter by a factor of two and a half times. The stimulation site is debated, and may be at the exact conjunction of the coils or at the anterior and posterior point of the coils' junction, but it produces a 1 cm to 2 cm spot of electricity in the brain that stimulates underneath the identified point of the coil junction [33, 34]. This probably stimulates the axons leaving the pyramidal cells at the first few internodes, where the rate of change of the electrical field with respect to the axon is maximal.

Several researchers have observed that movement of a butterfly coil from 1/2 to 2 cm can cause complete appearance and disappearance of the EMG response in a limb muscle such as APB [33-35]. This indicates that a small area of cortex is being activated and substantiates the mathematical models of the electrical fields induced by the stimulators. This degree of "resolution" over the cortex, 1-2 cm, is enough to identify functional cortex that is a critical structure needing protection next to a brain lesion.

The correspondence between the scalp site stimulated and the brain location responding to stimulation is critical. If there is an error in this match, it raises fears that unreliable or misleading information could be produced. We want to examine this issue. In the case of the butterfly coil, the field's penetration is weak and unlikely to be more than 1-2 cm into the cortex. Further, the design of the coil makes localization of the site of stimulation easier. In a worse-case scenario, the error in registering the site of the stimulation coil with respect to the scalp is about 1 cm, because the junction is about 2 cm long.

We used a system of placing capsules of a fat (Vitamin E) which is visible on the MRI onto the patient's head at regular 2-cm intervals. The scalp stimulation points are measured from the same capsules, and then they can be used as reference points on the MRI to scale the scalp stimulation sites for superimposition on the brain. This produces a map of EMG response in a selected muscle, such as APB in terms of brain sites stimulated. The map itself is shown in Figure 5, with a skyscraper plot of the amplitude of compound muscle action potentials in terms of scalp position for a given muscle. This superimposition process probably has an error of 1 cm, the width of each grid square. We estimate the maximum error in location to be 2 cm, and the likely error about 1 cm. This is tolerable but not the desired accuracy.

A more accurate map may be producible by some combination of an MRI scanner and a transcranial magnetic stimulation map, using the mapping coil to induce a defect in the MRI scan that indicates the site of the magnetic field and suggests the location of the induced electric field. This would be a precise correspondence of the anatomy and the physiology, but requires advanced integration of the electronics in the future.

Clinical cortical mapping where surgery is planned can have two major consequences, ie., goals for us to pursue. First, it can indicate that the functional cortex is so near a lesion that it may well be damaged in trying to reach the lesion, risking damage and forcing a careful planning so that the approach will miss the functional brain. Or it can show that there is not a major risk because the functional cortex is separated from the lesion, as long as the functional cortex is identified and avoided in approaching this lesion. Examples of the mapping method for neurosurgery have been published for each of these possibilities.

Our clinical experience has included the two illustrative cases we have reported, one for each category, that we wish to discuss here. The first case was a 47-year-old male with an onset of left arm weakness, who had a metastatic brain lesion in the right motor cortex. This lesion was mapped by the butterfly coil, and the areas of cortex that were activated by the stimulation lay at their closes 1 cm to 2 cm distant from the lesion. This provided a safe margin for surgical excision of the lesion. The surgical removal was successful without producing a deficit.

The second case is a 62-year-old male with a 10-year seizure history and a lesion in the right side of the brain. Cortical mapping with the butterfly coil indicated that there was motor-cortex activated by the stimulation overlying the posterior half of the tumor. However, the anterior half of the tumor showed no responses to stimulation. This, then, identifies a safe path for the approach to the lesion, and one that could not be identified by MRI scans. Also in this case, a needle biopsy of the lesion was done through the anterior aspect, without deficit.

DISCUSSION

These two cases illustrate what could be developed by a marriage of the magnetic stimulators used for cortical mapping and the imaging power of MRI scanners. The MRI scanner can show the anatomical locations over which the stimulation coil is positioned, and it can show the direction of the local anatomy. The local anatomy is important in better cortical mapping interpretation because of two factors. First, the direction of gyrus conformations influences the effectiveness of the magnetic stimulator. As the magnetic stimulator is most effective at points of maximal rate of change in the magnetic field, and its effectiveness falls off with distance, the local anatomy is very influential.

For example, cortical atrophy can affect the power required to reach the cortex. One consequence of the rate of change effect is that the point of bending of an axon can produce a high rate of change and therefore a point of stimulation as it curves out of the field. Thus the location of gyri with respect to the field is important, particularly for small fields such as the butterfly or figure-of-eight coil produces. The second major role of the MRI scanner is to identify the local contours of pathology. This can tell where the responding cortex is next to a lesion, as in our examples.

Is there evidence that the cortical maps can vary, enough to bring a solid clinical benefit to mapping? Levy and Amassian have reported that major rearrangements in the cortical area controlled by muscle occur after spinal cord injury, with substantial expansion of the cortical surface responding to stimulation for a muscle just above the level of the dysfunction [36]. Cohen and Hallet confirmed this finding [37-39]. The time course of this is not yet clear.

Furthermore the potential of neurological deficits if the responding cortex is injured in a surgical procedure or event of a disease is not known. There is evidence that this can occur in other conditions. Cohen et al. confirmed that these effects are present in patients with an amputated limb, and even without such a traumatic neurological change as paralysis or limb amputation, changes in cortical-map representation occur. Cohen et al. reported that with congenital mirror movements there is a "marked derangement of the map of outputs of distal hand muscles with enlarged and ipsilateral representations." Such changes suggest that with disease or injury a reliable tool for mapping of the cortex is needed which can allow an informed clinical plan to be made about the risk of a given surgical approach.

Further valuable linkages between MRI and magnetic stimulation include the use of the MRI scanner to image the stimulation field of the magnetic stimulator as a perturbation or hold in the MRI image. This would allow the identification of the stimulation site to be precise. Armed with this tool, stimulation of local areas and recording from other areas of the scalp could lead to more complete testing of electrophysiological capacity. Magnetic stimulation has strong contributions it can add to cortical mapping.

Ultimately the delivery of these capabilities, mapping or operative monitoring, which appear so promising in their research demonstrations, depend on two accomplishments. There is a consensus needed that the test is beneficial to the patient, enough to outweigh any identified risk. This means government regulatory approval, such as from the Food and Drug Administration in the United States. And it must make the transition to a clinical tool, routine in use, reliable in the production of test results, relatively widely available with some standard constructions. A clinically routine mapping device should test and deliver results in an hour or so. These two steps, in the United States at any rate, require substantial further development, which should be encouraged.

REFERENCES

1. A.T. Barker, R. Jalinous, and I.L. Freeston, Non-invasive magnetic stimulation of human motor cortex, *Lancet* 1:1106-1107 (1985).
2. A.T. Barker, J.L. Freeston, R. Jalinous, and J.A. Jarratt, Noninvasive stimulation of motor pathways within the brain using time-varying magnetic fields, *Electroencephalography & Clinical Neurophysiology* 61:245-246 (1985).
3. V.E. Amassian, G.J. Quirk, and M. Stewart, A comparison of corticospinal activation by magnetic coil and electrical stimulation of monkey motor cortex, *Electroencephalography & Clinical Neurophysiology* 77:390-401 (1990).
4. V.E. Amassian, R.Q. Cracco, and P.J. Maccabee, A sense of movement elicited in paralyzed distal arm by focal magnetic coil stimulation of human motor cortex, *Brain Research* 479:355-360 (1989).
5. V.E. Amassian, R.Q. Cracco, and P.J. Maccabee, Focal stimulation of human cerebral cortex with the magnetic coil: a comparison with electrical stimulation, *Electroencephalography & Clinical Neurophysiology* 74:401-416 (1989).
6. R.Q. Cracco, V.E. Amassian, P.J. Maccabee, and J.B. Cracco, Comparison of human transcallosal responses evoked by magnetic coil and electrical stimulation, *Electroencephalography & Clinical Neurophysiology* 74:417-424 (1989).

7. P.J. Maccabee, V.E. Amassian, R.Q. Cracco, and J.A. Cadwell, An analysis of peripheral motor nerve stimulation in humans using the magnetic coil, *Electroencephalography & Clinical Neurophysiology* 70:524-533 (1988).
8. V.E. Amassian, R.Q. Cracco, P.J. Maccabee, J.B. Cracco, A. Rudell, and L. Eberle, Suppression of visual perception by magnetic coil stimulation of human occipital cortex, *Electroencephalography & Clinical Neurophysiology* 74:458-462 (1989).
9. R.Q. Cracco, V.E. Amassian, P.J. Maccabee, and J.B. Cracco, Excitatory and inhibitory effects of magnetic coil stimulation of human cortex, *Electroencephalography & Clinical Neurophysiology - Supplement* 41:134-139 (1990).
10. P.J. Maccabee, L. Eberle, V.E. Amassian, R.Q. Cracco, A. Rudell, and M. Jayachandra, Spatial distribution of the electric field induced in volume by round and figure-eight magnetic coils: relevance to activation of sensory nerve fibers, *Electroencephalography & Clinical Neurophysiology* 76:131-141 (1990).
11. V.E. Amassian, P.J. Maccabee, and R.Q. Cracco, Focal stimulation of human peripheral nerve with the magnetic coil: a comparison with electrical stimulation, *Experimental Neurology* 103:282-289 (1989).
12. V.E. Amassian, R.Q. Cracco, and P.J. Maccabee, Focal magnetic coil activation of human motor cortex elicits a sense of movement in ischaemically paralysed, distal arm, *J Physiol (Lond)* 403:75 (1988).
13. V.E. Amassian, R.Q. Cracco, and P.J. Maccabee, Focal stimulation of human cerebral cortex with the magnetic coil: A comparison with electrical stimulation, *Electroencephalography & Clinical Neurophysiology* 74:401-416 (1989).
14. M. Machida, S.L. Weinstein, T. Yamada, J. Kimura, and S. Toriyama, Dissociation of muscle action potentials and spinal somatosensory evoked potentials after ischemic damage of spinal cord, *Spine* 13:1119-1124 (1988).
15. J.H. Owen, K.H. Bridwell, R. Grubb, A. Jenny, B. Allen, A.M. Padberg, and S.M. Shimon, The clinical application of neurogenic motor evoked potentials to monitor spinal cord function during surgery, *Spine* 16:S385-S390 (1991).
16. K. Tsuchida, A. Hashimoto, R. Seino, T. Hirayama, S. Aomi, and H. Koyanagi, Two case reports of neurologic complications after aortic aneurysm operation [Jpn], *Nippon Kyobu Geka Gakkai Zasshi - Journal of the Japanese Association for Thoracic Surgery* 37:1995-2000 (1989).
17. M.H. Zornow, M.R. Grafe, C. Tybor, and M.R. Swenson, Preservation of evoked potentials in a case of anterior spinal artery syndrome, *Electroencephalography & Clinical Neurophysiology* 77:137-139 (1990).
18. M.H. Zornow and J.C. Drummond, Intraoperative somatosensory evoked responses recorded during onset of the anterior spinal artery syndrome [see comments], *Journal of Clinical Monitoring* 5:243-245 (1989).
19. R.F. Ghaly, J.L. Stone, A. Aldrete, and W.J. Levy, Motor evoked potentials in primates; anesthetic considerations, *J Neuroanesthesia* 2000:182 (1991).
20. R.F. Ghaly, J.L. Stone, W.J. Levy, R. Kartha, and A. Aldrete, The effects of nitrous oxide on transcranial magnetic-induced electromyographic responses in the monkey, *J. Neurosurg Anesth* 2:175-181 (1990).
21. S.S. Haghighi, K.D. Green, J.J. Oro, R.K. Drake, and G.R. Kracke, Depressive effect of isoflurane anesthesia on motor evoked potentials, *Neurosurgery* 26:993-997 (1990).
22. S.S. Haghighi, R. Madsen, D. Green, J.J. Oro, and G. Kracke, Suppression of motor evoked potentials by inhalation anesthetics, *J Neurosurg Anesth* 2:75-78 (1990).
23. T.J. Lasasso, J.B. Boudreaux, D.A. Muzzi, R.F. Cuchiara, and J.R. Daube, The effect of anesthetic agents on transcranial magnetic motor evoked potentials (TMEP) in neurosurgical patients (Abstract), *J Neurosurg Anesth* 3:200 (1991).
24. T.B. Sloan and D. Levin, Effect of enflurane, halothane, isoflurane and nitrous oxide on cortical magnetic motor-evoked potentials (Abstract), *J Neurosurg Anesth* 3:201 (1991).
25. H.L. Edmonds, Jr., M.P. Paloheimo, M.H. Backman, J.R. Johnson, R.T. Holt, and C.B. Shields, Transcranial magnetic motor evoked potentials (tcMMEP) for functional monitoring of motor pathways during scoliosis surgery, *Spine* 14:683-686 (1989).
26. R.F. Ghaly, J.L. Stone, W.J. Levy, R. Kartha, and E.B. Brunner, The effect of an anesthetic induction dose of midazolam on motor evoked potentials evoked by transcranial magnetic stimulation in the monkey, *J Neuroanesthesia* 20:183 (1991).
27. R.F. Ghaly, J.L. Stone, W.J. Levy, R. Kartha, E.A. Brunner, R.A. Aldrete, and R. Laige, The effect of neuroleptanalgesia (Droperidol-Fentanyl) on motor potentials evoked by transcranial magnetic stimulation in the monkey, *J Neurosurg Anesth* 3:117-123 (1991).
28. R.F. Ghaly, J.L. Stone, W.J. Levy, P. Roccaforte, and E.B. Brunner, The effect of etomidate on transcranial magnetic-induced motor evoked potentials in the monkey, *Neurosurgery* 20:184 (1990).
29. R.F. Ghaly, J.L. Stone, A. Aldrete, and W.J. Levy, Effects of incremental ketamine hydrochloride doses on motor evoked potentials (MEPs) following transcranial magnetic stimulation: A primate study, *J Neurosurg Anesth* 2:79-85 (1990).

30. R.F. Ghaly, J.L. Stone, W.J. Levy, R. Kartha, M.L. Miles, and H.J. Jaster, The effect of etomidate or midazolam hypnotic doses on motor evoked potentials in the monkey, *J Neurosurg Anesth* 3:20-27 (1991).
31. T.B. Sloan, Vecuronium alters cortical magnetic motor evoked potentials, *J Neurosurg Anesth* 2:251 (1990).
32. S. Ueno, T. Tashiro and K. Harada, Localized stimulation of neural tissues in the brain by means of a paired configuration of time-varying magnetic fields, *J Appl Phys* 64: 5862-5864 (1988).
33. V.E. Amassian, R.Q. Cracco, P.J. Maccabee, B. Bigland-Ritchie, and J.B. Cracco, Matching focal and non-focal magnetic coil stimulation to properties of human nervous system: mapping motor unit fields in motor cortex contrasted with altering sequential digit movements by premoter-SMA stimulation, *Electroencephalography & Clinical Neurophysiology - Supplement* 43:3-28 (1991).
34. W.J. Levy, V.E. Amassian, U.D. Schmid, and C. Jungreis, Mapping of motor cortex gyral sites non-invasively by transcranial magnetic stimulation in normal subjects and patients, *Electroencephalography & Clinical Neurophysiology - Supplement* 43:51-75 (1991).
35. B.U. Meyer, R. Diehl, H. Steinmetz, T.C. Britton, and R. Benecke, Magnetic stimuli applied over motor and visual cortex: influence of coil position and field polarity on motor responses, phosphenes, and eye movements, *Electroencephalography & Clinical Neurophysiology -Supplement* 43:121-134 (1991).
36. W.J. Levy, Jr., V.E. Amassian, M. Traad, and J. Cadwell, Focal magnetic coil stimulation reveals motor cortical system reorganized in humans after traumatic quadriplcgia, *Brain Research* 510:130-134 (1990).
37. L.G. Cohen, B.J. Roth, E.M. Wassermann, H. Topka, P. Fuhr, J. Schultz, and M. Hallett, Magnetic stimulation of the human cerebral cortex, an indicator of reorganization in motor pathways in certain pathological conditions, *Journal of Clinical Neurophysiology* 8:56-65 (1991).
38. L.G. Cohen and M. Hallett, Noninvasive mapping of human motor cortex, *Neurology* 38:904-909 (1988).
39. L.G. Cohen, S. Bandinelli, H.R. Topka, P. Fuhr, B.J. Roth, and M. Hallett, Topographic maps of human motor cortex in normal and pathological conditions: mirror movements, amputations and spinal cord injuries, *Electroencephalography & Clinical Neurophysiology - Supplement* 43:36-50 (1991).

DISCUSSION

Chairman Prof. S. Matsuoka: Thank you very much, Dr. Kraus. Are there any questions or comments? Yes, please.

Dr. M. Fujiki: You have successfully recorded motor-evoked potentials by using the cap coil as a stimulator during general anesthetic operating conditions. Are the resulting wave forms as reliable and stable as spinal evoked potentials? Also, have you recorded spinal motor evoked potentials simultaneously, and if so, did you see "D" and "I" waves?

Dr. Kraus: Operative monitoring using transcranial magnetic stimulation recorded from distal muscles is more difficult than spinal evoked potentials because of the effects of anesthetics. The anesthetic effects and muscle relaxant effects also result in more variability than spinal evoked potentials.

If these factors can be successfully controlled, however, transcranial magnetic stimulated motor evoked potentials have several advantages over spinal evoked potentials. First, there is the advantage of monitoring before surgery and during patient positioning. Second, the pathways must be motor and must cross gray-matter synapses. Spinal evoked potentials only monitor white-matter pathways and these may be motor or antidromally conducted sensory pathways. Our technique can be performed while the patient is awake with descending motor pathways, and the monitored pathways must include spinal cord gray matter which is most susceptible to ischemia. The latter is especially important in monitoring aneurism surgeries. We have not recorded spinal evoked potentials simultaneously. Since we record compound muscle action potentials, "D" and "I" waves cannot be identified.

Dr. Y. Mano: You said that motor representations change after spinal cord injury and gave three possible explanations. We have also seen this phenomenon, but in our experience, the amplitude of the induced motor evoked potentials of the maps decreases as the motor representation area increases. Your presentation did not suggest decreases in amplitudes at all. Could this be an effect of the location of the patient's injury?

Dr. Kraus: I do not know if the amplitude of the evoked potentials decreases, as these were clinical cases where mapping of the motor cortex prior to injury was impossible. Both patients presented here had lower cervical lesions, specifically, C4 - 5.

Prof. Matsuoka: Any other questions? Then, thank you very much, Dr. Kraus.

MAGNETIC STIMULATION OF THE HEART AND SAFETY ISSUES IN MAGNETIC RESONANCE IMAGING

John Nyenhuis, Joe Bourland, Gabriel Mouchawar, Leslie Geddes, Kirk Foster, Jim Jones, William Schoenlein, George Graber, and Tarek Elabbady

Hillenbrand Biomedical Engineering Center
Purdue University
West Lafayette, Indiana 47907 USA

D. Joseph Schaefer and Mark Riehl

General Electric Medical Systems
Waukesha, Wisconsin 53188 USA

1.0 INTRODUCTION

Our group at Purdue University has been studying the physiological effects of pulsed magnetic fields for several years. The initial work was directed toward cardiac pacing with a pulsed magnetic field. Our motivation was the development of a non-invasive and relatively pain-free method for cardiac stimulation. We were the first to induce cardiac ectopic beats with pulsed magnetic fields in the closed-chest dog [1,2]. Unfortunately, the large energies required to stimulate the heart preclude the development of a portable magnetic cardiac pacemaker.

Magnetic stimulators used medically generate intense rapidly changing magnetic fields capable of stimulating nerves [3]. Fast magnetic resonance image (MRI) techniques such as echo planar imaging offer high patient throughput and reduce motion artifacts [4]. Pulsed magnetic fields used in prototype imagers are of sufficient intensity to elicit sensation in human volunteers [5]. The risk of producing cardiac arrhythmias by these devices is of concern. In addition to the pulsed magnetic fields from the gradient coils, the dc magnetic field and the radio frequency (rf) field are present in MRI imagers.

In this paper, we first discuss our work on cardiac stimulation in dogs with pulsed magnetic fields. We then discuss the safety considerations involved with the electromagnetic fields present in MRI imagers. We report our measurement of cardiac thresholds from pulsed magnetic fields in MRI z-gradient coils. Lastly, we present results on human peripheral nerve stimulation from z-gradients. An important conclusion is that cardiac stimulation requires significantly larger pulsed field intensity than does peripheral nerve stimulation.

2.0 CLOSED-CHEST CARDIAC STIMULATION WITH A PULSED MAGNETIC FIELD

2.1 Motivation

At an international conference on motor evoked potentials held at Purdue University in 1986 [6], Tony Barker demonstrated the magnetic stimulator developed at the University of Sheffield [3]. Well-known advantages of magnetic stimulation of the brain over stimulation with electrodes include non-invasiveness and greatly reduced pain. Two of us (Geddes and Bourland) hypothesized that magnetic stimulation of the heart would have advantages over the currently used techniques, which utilize electrodes. The original Purdue concept for cardiac pacing is shown in Figure 1. Attending medical personnel would put a coil over the patient's heart, and the power supply would provide the energy needed to produce the magnetic field pulses to stimulate the heart. During our work, we demonstrated what huge amounts of energy are required to stimulate the heart; the power supply for a magnetic cardiac pacer would need to be much larger than shown in Figure 1.

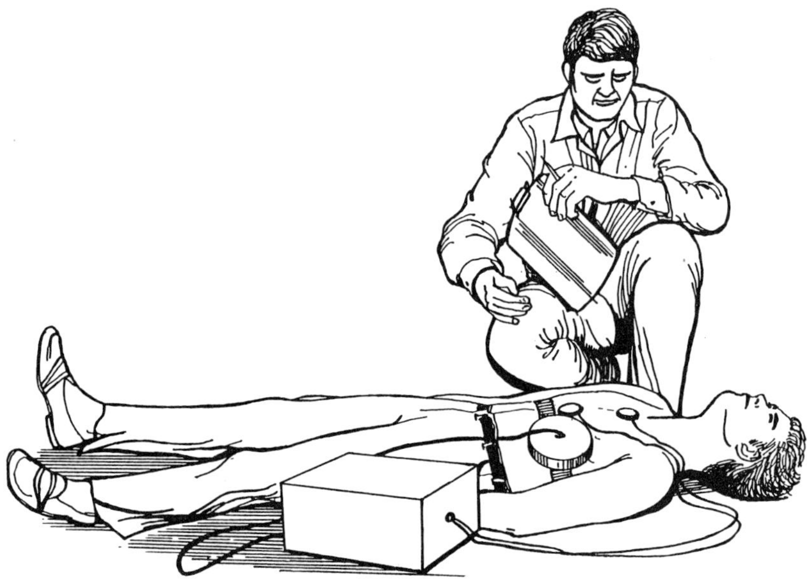

Figure 1. A patient receiving emergency cardiac pacing with a pulsed magnetic field.

The most common method for temporary cardiac pacing is invasive and requires passing a pacing lead through the superior or inferior vena cava to the heart. The heart can be paced non-invasively by placing electrodes on the chest, and semi-invasively by passing a pacing lead via the esophagus. While placement of electrodes on the chest is non-invasive, it is painful because current to the heart must pass through the nearly insulating skin, through the bones of the thorax, and the chest muscles. Proper placement of an esophageal pacing lead can be tedious and time consuming, and requires skill. We at Purdue hoped to devise an improved method of non-invasive cardiac pacing by using a pulsed magnetic field.

Figure 2 depicts the mechanism of cardiac pacing by a pulsed magnetic field. The heart is exposed to the magnetic field from the coil, which is connected to a pulse generator. A pulse of current in the coil generates a time-varying magnetic field, which induces eddy currents in the heart muscle. If the induced eddy currents are of sufficient intensity and of adequate duration, cardiac pacing will result.

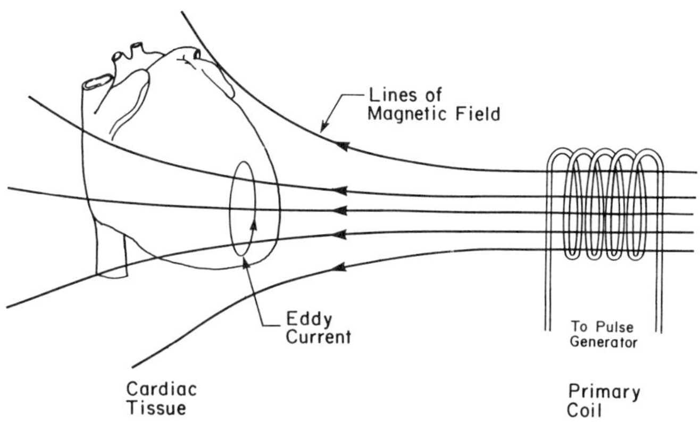

Figure 2. Mechanism for stimulation of the heart with a pulsed magnetic field.

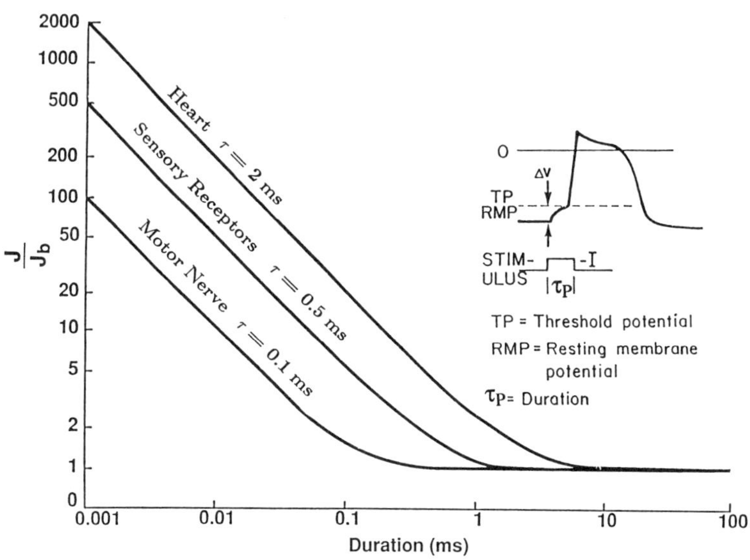

Figure 3. Strength-duration curves for motor nerve, sensory receptors and cardiac muscle. J is the current density required for stimulation and J_b is the rheobase current density. The normalized current density J/J_b is plotted vs. duration τ_p of a rectangular stimulus waveform. The inset shows the time dependence of the membrane potential during stimulation.

2.2 Theory

For a rectangular stimulating pulse of duration τ_p, the Blair expression [7,8] for the strength duration curve in terms of current density J is:

$$J = \frac{J_b}{1 - e^{-\tau_p/\tau}}$$

where J_b is rheobase current density and τ is the membrane time constant, a quantity that is specific for each type of tissue. Figure 3 shows intensities of rectangular pulses normalized to rheobase versus stimulus duration for several tissues. For motor nerve, a typical τ is 100 μs and sensory receptors have a somewhat longer time constant, about 500 μs. Our target organ, the heart, has a much longer membrane time constant, approximately 2 ms. Reilly predicts that rheobasic current densities for motor nerve and the heart are approximately the same, based on a review of the literature and computer models [9].

In magnetic stimulation, it is not energetically practical to have a very long duration pulse. For stimulation of nerve, pulse durations between 50 μs and 100 μs are typically used. In magnetic stimulation, the induced electric field is believed to be proportional to the time-varying magnetic field dB/dt. For $\tau_p << \tau$ and identical J_b, the peak magnetic field B in the coil to achieve stimulation is proportional to τ. Since the time constant for cardiac muscle is as much as 20 times greater than the time constant for peripheral nerve, the peak magnetic field, B_{peak} for cardiac stimulation, may be 20 times that required for stimulation of peripheral nerve. The peak magnetic field energy delivered to the coils, which is proportional to B^2_{peak}, is thus expected to require up to 400 times as much energy as magnetic stimulation of peripheral nerve. These energy expectations guided our design of the stimulator and coils.

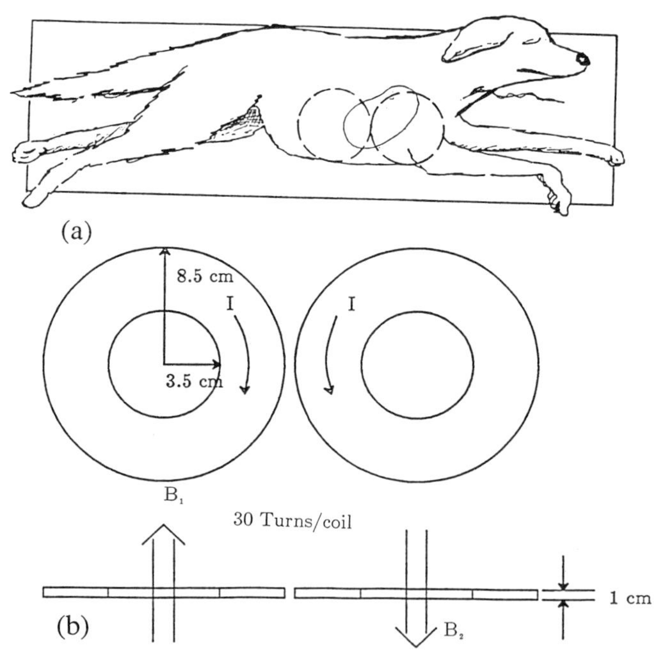

Figure 4. (a) Position of the co-planar coils (broken circles) under the left side of the dog; (b) dimensions and current flow in the coils.

2.3 Materials and Methods

In our experiments, the time-varying magnetic field is generated by discharging a capacitor bank into a coil. A simplified equivalent circuit for the magnetic stimulator and coils is a series RLC circuit, comprising an 820 μF, 9000-volt capacitor, output switch (ignitron), and typical coil inductance of 200 μH and total series resistance of 100 mΩ. A 220 mΩ resistance and reverse biased diode is connected across the 820 μF capacitor to limit reverse polarization of the capacitors.

For the study of cardiac dB/dt threshold in the dog, a pair of co-planar coils (see Figure 4), as proposed by Ueno [10, 11], was placed under the chest of the dog. The magnetic fields above the centers of the two coils are in opposing directions. In addition to providing a focused field [11], the co-planar pair has been found to be energetically efficient [12]. Each turn consisted of 30 turns of copper strip with height 1.27 cm, inner diameter 7 cm and outer diameter 17 cm. The coils were mounted with their edges touching; this geometry provides optimum stimulation of a target 4 cm from the adjacent coil edges [12].

Figure 5 shows the calculated current density in a rectangular prism with uniform conductivity below a co-planar pair of coils. The induced current is greatest between the two coils, zero above the center of the coils, and reverses direction at the edge of the conductor. We placed the heart of the dog above the point at which the coils touch. For the cardiac stimulation work, we were not particularly interested in focal stimulation, because the heart is a relatively large organ. We were most interested in minimizing the energy required for stimulation [12].

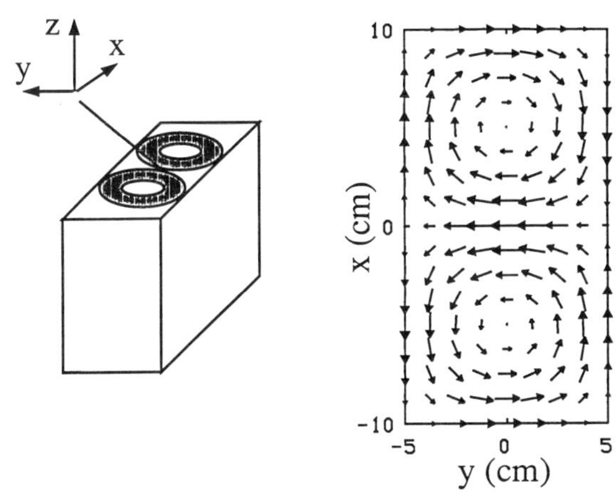

Figure 5. Co-planar coil pair above a rectangular conducting prism and the calculated eddy current flow.

The reason the co-planar pair can be more energetically efficient can be explained as follows. The two coils are placed next to each other. For a given amount of time-varying current, dI/dt, in each coil, placing two coils together results in twice as much electric field as would be produced by one coil. The inductance of the two coils is approximately twice that of the single coil, because the coupling is small. For a single coil, doubling the induced electric field results in a quadrupling of the peak energy in the magnetic field, $1/2\ LI^2_{max}$. Doubling the induced electric field by using two coils results in a doubling of the peak magnetic field energy. The conclusion is that the two-coil geometry can be nearly twice as efficient energetically as a single coil for stimulation purposes [12].

In addition, we calculated that for energetically efficient stimulation, the coil outer radius should be about twice the target depth. The target for cardiac stimulation is about 4 cm, resulting in an outer radius of 8 cm for a coil with optimal energy efficiency. To maximize the magnetic field energy, the coil height should be as small as possible. However, a thin coil results in excessive Joulean heating. The 1.27 cm coil height was felt to be a good compromise that resulted in acceptable magnetic field energy and Joulean heating.

As Barker has shown [13], the stimulus duration can be chosen to minimize the energy in the pulse generator required to achieve a given level of stimulation. Figure 6 shows the calculated stored energy in the capacitor versus pulse duration for stimulation of the canine heart. Rheobasic current density of 1.48 mA/cm^2 reported by Pearce et al. [14] and myocardial resistivity of 400 Ω-cm reported by Geddes and Baker [15] were used in the calculations. For long duration, the stored energy becomes very large. For durations less than 1 ms, the stored energy is approximately independent of the duration. In order to keep voltage, current and inductance within practical limits, a duration of about 600 μs was chosen for the cardiac stimulation attempts. This pulse duration was achieved by matching coil inductance to the capacitance of the generator.

The coils are wrapped in glass epoxy to prevent their failure from the large Lorentzian forces. The acoustic report from the coils was significant at the higher energy levels, although no attempt was made to quantify the sound from the coils.

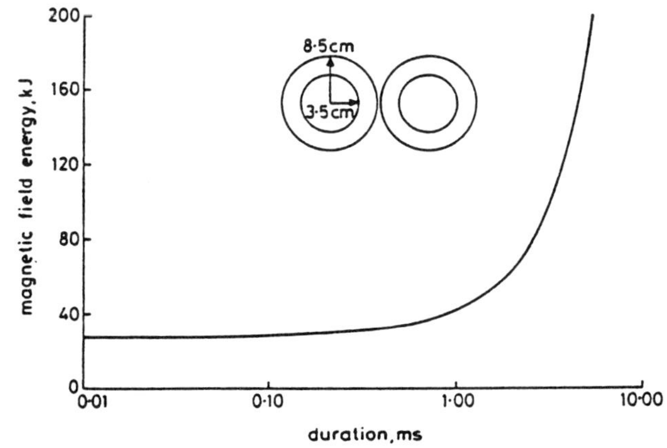

Figure 6. Predicted energy required vs. stimulus duration for stimulating cardiac muscle with a pulsed magnetic field.

The calculated voltage and current waveforms in the coil are shown in Figure 7. The peak in current, shown in Figure 7(a), occurs at 600 μs. The dB/dt waveform, shown in Figure 7(b), is proportional to the derivative of the current waveform. The time between the onset and the first zero-crossing is defined as the duration τ_p. The negative part of the dB/dt waveform is due to the series diode and resistor, which are in parallel with the capacitor. The shunting diode is used to minimize reverse voltage across the electrolytic capacitors.

2.4 Animal Preparation

All experiments were performed in accordance with the Guide for the Care and Use of Laboratory Animals, published by the U.S. Department of Health, Education, and Welfare (DHHS publication No. NIH 85-23, revised 1985). In addition, these studies were approved by the Purdue University Animal Care and Use Committee.

For the study, 11 mongrel dogs of both sexes, with body weight from 17 to 26 kg, were anesthetized with sodium pentobarbital (30 mg/kg, i.v.), intubated and ventilated with room air. A lead II ECG and direct arterial pressure were recorded. Temporary and reversible cardiac arrest was induced by electrical stimulation of the right vagus nerve. A test magnetic pulse was delivered 3 seconds after initiation of vagal stimulation, and a successfully induced ectopic beat was noted by a QRS complex in the ECG and a pulse in arterial pressure. The co-planar coil pair was placed below the heart of the dog as depicted in Figure 4.

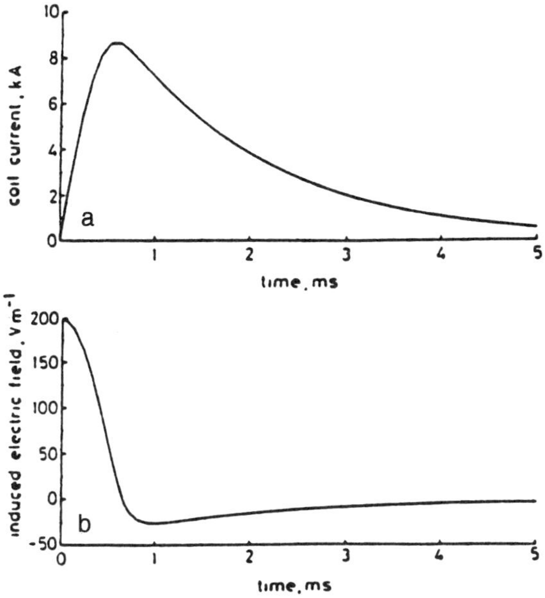

Figure 7. Calculated current and induced electric field waveforms for closed chest cardiac stimulation with a pulsed magnetic field.

2.5 Results

Evidence of cardiac stimulation by a pulsed magnetic field in a typical subject is presented in Figure 8. Temporary cardiac arrest can be appreciated by noting cessation of electrical activity from the heart in the ECG, as in Figure 8(a), and loss of arterial pulses in the pressure channel, as in Figure 8(b). Artifact from the magnetic pulse is evident in both channels. Shortly after delivery of the magnetic pulse, depolarization of the ventricles at an ectopic site near the ventricular apex can be noted in the ECG, followed by an arterial pulse in the pressure channel. Normal sinus rhythm returns shortly after termination of vagal stimulation.

The stimulation threshold was determined by applying magnetic field pulses of different intensity. The cardiac threshold was defined as the lowest voltage that would cause stimulation, but which was no more than 10 percent greater than a pulse which would not cause stimulation.

A pulse generator energy of 11,800 ± 1800 J (mean ± standard deviation) was required to achieve closed-chest, magnetically induced ectopic beats with the co-planar coils in 11 dogs. This is much larger than the few hundred Joules of energy per pulse that are produced in a magnetic stimulator for the brain.

For the co-planar coil and assuming that charge accumulated at boundaries between tissues with different resistivity is insignificant, the mean peak induced field for threshold stimulation in the dog was 213 V/m for the 571 μs damped sine wave pulse. Using the Blair expression and membrane time constant $\tau = 2.14$ ms, rheobasic field strength for the damped sine wave is 49.8 V/m, which corresponds to rheobase for a rectangular pulse of 28.9 V/m.

It is unclear why the rheobasic electric field strength calculated from the measurements of this study are approximately one-half of those based on the data from Pearce et al.[14]. Nonetheless, the stimulation cardiac thresholds are sufficiently large that a magnetic cardiac pacer would require considerable power.

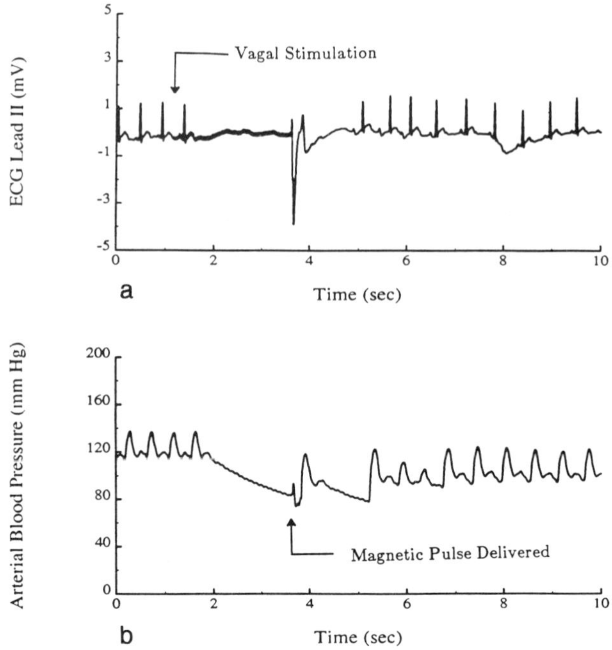

Figure 8. ECG (a) and blood pressure (b) waveforms observed for stimulation of the canine heart with a pulsed magnetic field. During cardiac arrest (induced by vagal stimulation) the magnetic stimulator evoked an ectopic beat, shown by the QS-wave and the blood-pressure pulse.

3.0 SAFETY ISSUES IN MAGNETIC RESONANCE IMAGING

3.1 The MRI Physical Environment

Figure 9 depicts a patient in a Magnetic Resonance (MR) imager. The patient undergoing an MRI scan is exposed to three types of electromagnetic fields.

(i) The static field B_o. A key component of MR imagers is the highly uniform static magnetic field B_o which typically ranges from 0.5 to 2.0 T; experimental imagers utilizing static fields as large as 4.0 T have been reported. The magnet is usually superconducting, although resistive and permanent magnets have been used. Larger field strengths result in improved resolution, but also increased costs and imager size. The static field is customarily defined to be along the z-coordinate direction.

(ii) The rf field. The rf field is used to induce precession in the nuclei of interest. For protons, the rf frequency is 42.58 MHz/T and thus an rf frequency of 63.86 MHz is required for $B_o = 1.5$ T. Typical rf pulse widths and amplitudes are about 1 ms and about 0.1 Gauss.

(iii) The gradient fields. Gradient coils are used in MR imagers to (a) spatially select the portion of the patient in which the protons will be in resonance (slice select), (b) spatially alter the phase relative to some reference (phase encode), and (c) spatially vary the frequency of the protons while the time-varying flux is being measured by the receiver coil (frequency encode). Three pairs of gradient coils are present in the typical imager to produce

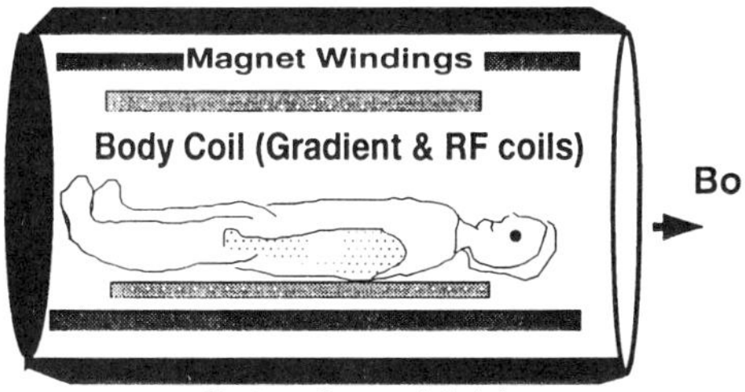

Figure 9. Patient in an MR imager and coils which produce the static, rf, and gradient magnetic fields.

field gradients: $\partial B_z/\partial x$, $\partial B_z/\partial y$, and $\partial Bz/\partial z$. Typical rates of change are less than 20 T/s and typical maximum field strengths are less than 40 G.

3.2 Safety Issues with MRI Fields

(i) The static field B_o. The static field is apparently well-understood and there are no known risks of exposure to fields of strengths now used in MR imagers. Potential bioeffects include the voltage induced transverse to flowing blood. This results in a T-wave artifact in the ECG. Discomfort has been reported upon considerable head movement in a 4-T field. The ultimate limitation on the static field is likely to be the economics of building and shielding huge field magnets.

(ii) The rf field. The principal bioeffect of the rf field is inductive heating in the subject [16]. Extensive work has been done on this topic and the heating effects are well understood [17]. A technical problem is that the power deposited by an rf pulse is proportional to the square of the static field strength. For instance, a 90° rf pulse at $B_o = 4$ T will result in about seven times as much energy deposition as would a 90° pulse at $B_o = 1.5$ T. The greater power deposition restricts the number of rf pulses that can be applied at higher static fields. Clinically, care must be taken to avoid rf burns from metallic objects on or within the patient [18].

(iii) The gradient fields. Gradient field frequencies are too low to cause significant heating in the patient. The principal potential bioeffect is stimulation of nerves and muscle by the induced eddy currents. Cohen et al. [5] reported sensation in human volunteers with dB/dt = 61 T/s rms from transverse coils. This field strength is consistent with estimates made by Reilly [19].

Our work at Purdue focused on the bioeffects of the gradient fields. In the following sections we describe our studies of cardiac stimulation in the dog and peripheral nerve stimulation in the human by z-gradient fields.

3.3 Cardiac Stimulation in the Dog with z-Gradients

The z-gradient $\partial B_z/\partial z$ in an imager with a longitudinal field is produced with a coil configuration which is essentially a Maxwell pair. The coil that we fabricated to simulate the stimulating effects of z-gradients is essentially 1/2 of the Maxwell pair in the z-coil. The geometry is depicted in Figure 10. The coil consists of 26 turns with width 13 cm on a cylindrical form with radius 13 cm. The radius is a factor 2.5 less than the coil radius that would be used for imaging humans. The smaller coil radius and using just one of the two coils of the Maxwell pair allowed obtaining the maximal peak magnetic field for the energy available in the generator.

The generator used is the one described in section 2.3. The capacitance of 820 μF resulted in a duration of 530 μs. The shape of the dB/dt waveform is shown for the MRI z-coil is shown in Figure 10. The generator provided a maximal magnetic field of 19 kJ.

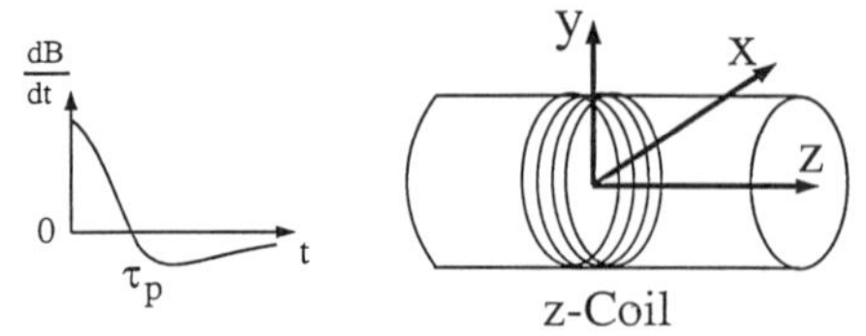

Figure 10. Model z-gradient coil for measuring cardiac stimulation thresholds in the dog and the dB/dt waveform used in the experiments. Coil parameters are given in the text and τ_p = 540 μs.

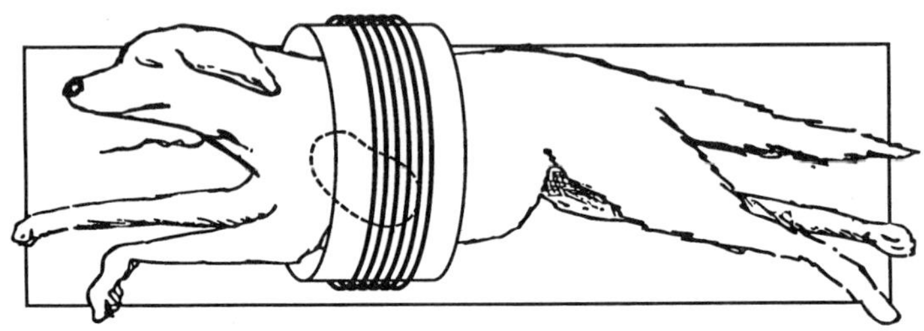

Figure 11. Placement of dog in MRI z-gradient coil for measurement of cardiac thresholds.

Animal preparation was the same as described in section 2.4 for cardiac stimulation with the co-planar pair. The dog was placed in the coil assembly as depicted in Figure 11with the cardiac apex at the z-axis center of the coil. Cardiac stimulation was attempted for three positions of the dog in the coil assembly: (a) dorsal, (b) left lateral and (c) right lateral. The heart was temporarily arrested by electrical stimulation of the right vagus nerve and current pulses were applied to the coil. Cardiac threshold was defined as the lowest intensity of magnetic field pulse that would produce stimulation, and it was established by varying the field in steps of less than 10 percent as described in section 2.5.

Results of the cardiac threshold measurements for the dog in the MRI z-gradient coil are shown in Figure 12. The symbols represent peak dB/dt (measured in the center of the coil assembly) cardiac stimulation thresholds for the waveform shown in Figure 10. Of the 10 dogs studied, eight could be stimulated within the generator limit of 5300 T/s in the right lateral position. Stimulation threshold for these eight dogs ranged from 3500 to 5000 T/s. Six of the dogs could be stimulated in the left lateral position; only three of the 10 dogs could be stimulated in the dorsal position.

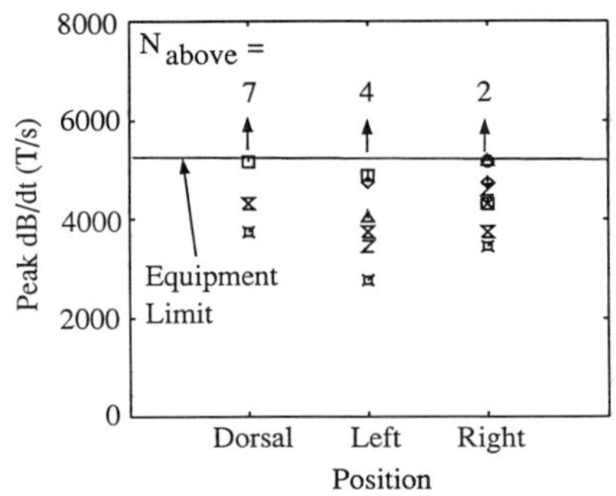

Figure 12. Cardiac stimulation thresholds for three positions of the dog in the MRI z-gradient coil.

It is interesting that more dogs could be stimulated in the right lateral than in the left lateral position. One might expect stimulation to be easier in the left lateral position because the heart of the dog is closer to the coils.

We emphasize that the measured field gradients for cardiac stimulation are much larger than gradients which are now used or which will be used in future imagers. Our experimental results indicate that it is highly unlikely that cardiac arrhythmias would be produced in an MR imager.*

3.4 Study of Human Peripheral Nerve Stimulation from z-Gradients

A study was performed to determine the threshold for peripheral nerve stimulation in humans by MRI z-gradients [22]. The coil assembly used is an unshielded GE Signal coil assembly. The z-windings are concentrated at the ends of the coil, where the z-component of magnetic field is maximal. The length of coil assembly is approximately 1 m and the radius is 33 cm. There was no static magnetic field present in these experiments. The calculated magnetic field for the z-coil is shown in Figure 13.

The simulated gradient pulses were obtained by discharging a capacitor bank into the coil. The equivalent circuit is shown in Figure 14. The z-coil is modeled by an inductance of 1.158 mH and a series resistance of 0.665 Ω. The generator capacitance and series resistance are represented by C and R_{cap}, respectively, in Figure 14. A series diode and resistance pro-

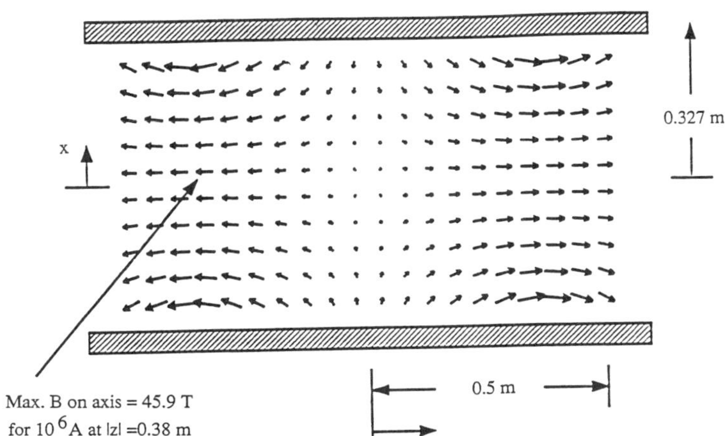

Figure 13. Map of the magnetic field for the z-gradient coil in the GE Signa whole body coil assembly.

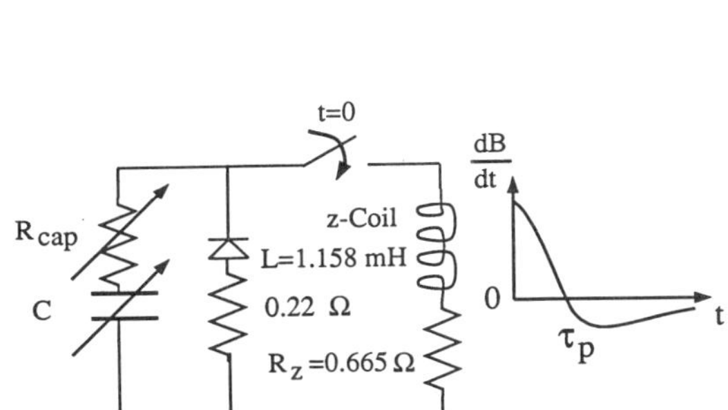

Figure 14. Equivalent circuit and dB/dt waveform for measurement of human peripheral nerve stimulation from z-gradients.

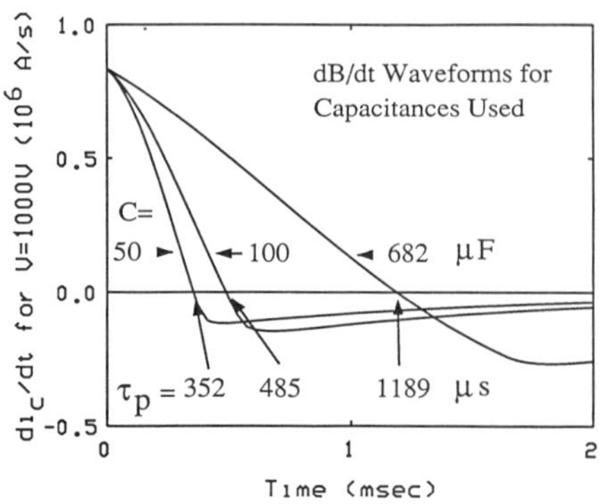

Figure 15. dB/dt waveforms used in measurement of human peripheral nerve threshold in the MRI z-gradient coil.

tect the capacitors from large reverse voltage. Calculated dB/dt waveforms are shown in Figure 15. The pulse duration τ_p is again defined as the zero crossing of the dB/dt waveform. Use of two different pulse generators, one of which has variable capacitance, resulted in pulse durations of 1189 μs, 485 μs, 352 μs, and 228 μs. The coil assembly has a voltage limit of 3000 V, resulting in a maximal dB_z/dt of 138 T/s.

Approval for the experimental protocol was obtained from the Purdue Committee on the Use of Human Subjects. Each volunteer was placed supine on a patient table with feet toward the coil. The subject was then advanced into the coil assembly and positioned so that the buttocks were in the region where the z-component of the time-varying magnetic field is greatest.

Although we expect that peripheral nerve threshold in the chest may be slightly lower than that in the pelvis, the study was conducted with the chest beyond the peak magnetic field because of concern for inducing cardiac arrhythmias. The measurements of cardiac stimulation thresholds, which were performed after the human studies, show that the risk of cardiac stimulation by pulsed MRI gradients is small. We are now confident that studies can be safely conducted to determine peripheral nerve threshold in the human chest.

Twenty healthy volunteers, 18 males and two females, participated in the study. Age range was from 23 to 68 years. Single capacitor discharges into an MRI z-gradient coil were used to induce stimulation in the pelvic region of the subjects. The intensity of the pulsed magnetic field was increased until the subject reported sensation. The most common description reported by the human volunteers was a light tap to the buttocks. Some subjects reported a slight tingling quality to the sensation; all subjects refuted pain as an apt descriptor of the sensation. A few subjects reported perceived stimulation in other locations, including the thigh, groin, and heel of the foot. Three of the 20 subjects felt nothing.

Raw data from the experiments are shown in Figure 16. Each circle represents the peak dB/dt that was threshold for a subject at the indicated stimulus duration. The dashed line represents the maximal dB/dt tested. The data follow a classical strength duration curve, with short-duration pulses requiring more intensity than longer ones. Not all subjects could be stimulated within the equipment limit. For the 1189 μs duration pulse, 17 of 20 subjects reported sensation. At this duration, the stimulation threshold ranged from 50 T/s to 120 T/s. This considerable variation may be due to subjective factors. For the 485 μs duration, 13 of 20 volunteers could be stimulated, at τ_p = 352μs, seven of 20 could be stimulated, and only one volunteer (the speaker!) reported sensation at τ_p = 228 μs.

It was attempted to correlate human peripheral nerve threshold to characteristics of the subjects such as weight or age. No significant correlation between stimulation threshold and age or weight was found.

In a recent work on stimulation by MRI gradient fields, Yamagata et al. compared stimulating effects of sinusoidal and trapezoidal echo-planar dB/dt waveforms [23].

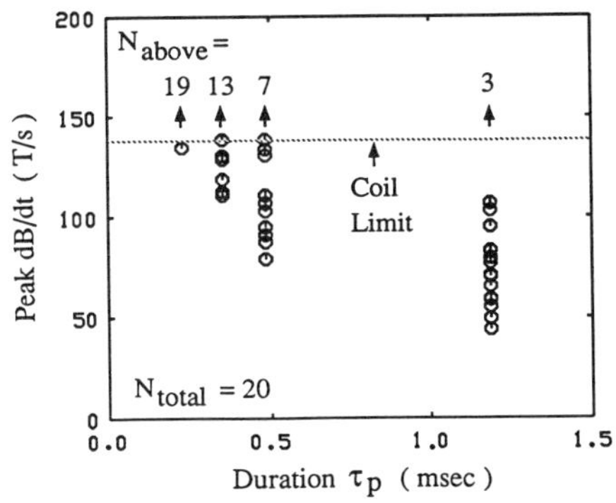

Figure 16. Peak dB/dt vs. duration τ_p required to elicit sensation in human volunteers in MRI z-gradient coil.

4.0 DISCUSSION AND CONCLUSIONS

In comparing Figures 12 and 16, we note that cardiac stimulation in the dog at a duration of 530 μs requires an dB/dt intensity approximately 30 times as great as does human peripheral nerve. When accounting for body geometry, this is consistent with our estimate that cardiac stimulation requires approximately 20 times the field intensity that is required for peripheral nerve stimulation in the human. Having established this safety margin, human peripheral nerve thresholds in the thoracic region can now be safely measured.

It should be noted that the magnetic field intensities employed in these studies are much greater than those used in current imagers. The goal of our work is to identify the safe limits for future MRI.

*Subsequent to the presentation at Kyushu University, we published additional information on stimulation of dogs in MRI gradient coils [20,21].

5.0 ACKNOWLEDGEMENTS

The speaker (JAN) wishes to thank Dr. Ueno for his considerable efforts to organize the International Symposium on Biomagnetic Stimulation. This work was funded in part by Fetzer Foundation, Kalamazoo, Michigan USA and by General Electric Medical Systems, Waukesha, Wisconsin USA.

REFERENCES

1. J.D. Bourland, G.A. Mouchawar, J.A. Nyenhuis, L.A. Geddes, K.S. Foster, J.T. Jones, and G.P. Graber, Transchest magnetic (eddy-current) stimulation of the dog heart, *Med. & Biol. Eng. & Computing* 28: 196-198 (1990).
2. J.D. Bourland, G.A. Mouchawar, J.A. Nyenhuis, L.A. Geddes, K.S. Foster, J.T. Jones and G.P. Graber, Closed-chest cardiac stimulation with a pulsed magnetic field, *Med. & Biol. Eng. & Computing* 30: 162-168 (1992).
3. A.T. Barker, I.L. Freeston, R. Jalinous and J.A. Jarrat, Magnetic stimulation of the human brain and peripheral nervous system: An introduction and the results of an initial clinical evaluation, *Neurosurgery* 20: 100-109 (1987).
4. P. Mansfield, Multi-planar image formation using NMR spin echoes, *J. Phys. C* 10: L55-L58 (1977).
5. M.S. Cohen, R.M. Weiskoff, R.R. Rzedian and H.L. Kantor, Sensory stimulation by time-varying magnetic fields, *Mag. Res. Med.* 14: 409-414 (1990).
6. *Neurosurgery,* vol. 20, (1987).

7. H.A. Blair, On the intensity-time relations for stimulation by electric currents -- I, *J. Gen. Physiol.* 15: 709-729 (1932).
8. H.A. Blair, On the intensity-time relations for stimulation by electric currents -- II, *J. Gen. Physiol.* 15: 731-755 (1932).
9. J.P. Reilly, Peripheral nerve and cardiac excitation by time-varying magnetic fields: A comparison of thresholds, *Med. & Biol. Eng. & Comput.* 29: 571-579 (1991).
10. S. Ueno, "Digests of the Intermag Conference," St. Paul, Minnesota, (1985).
11. S. Ueno, T. Tashiro and K. Harada, Localized stimulation of neural tissues in the brain by means of a paired configuration of time-varying magnetic fields, *J. Appl. Phys.* 64: 5862-5864 (1988).
12. J.A. Nyenhuis, G.A. Mouchawar, J.D. Bourland and L.A. Geddes, Energy considerations in the magnetic (eddy-current) stimulation of tissues, *IEEE Trans. Magn. MAG* 27: 680-687 (1991).
13. A.T. Barker, C.W. Garnham and I.L. Freeston, *in:* "Magnetic Nerve Stimulation: The Effect of Waveform on Efficiency," Motor Evoked Potentials Conference, August, Chicago, IL (1989).
14. J.A. Pearce, J.D. Bourland, W. Neilsen, L.A. Geddes, and M. Voelz, Myocardial stimulation with ul trashort duration current pulses, *PACE* 5: 52-58 (1982).
15. L.A. Geddes and L.E. Baker, "Principles of Applied Biomedical Instrumentation," Third Ed., Wiley, New York, NY (1989).
16. D.J. Schaefer, Safety aspects of magnetic resonance imaging, *in:* "Biomedical Magnetic Resonance Imaging: Principles, Methodology and Applications," F.W. Wehrli, D. Shaw and J. B. Kneeland, eds. 13, VCH Publishers, New York, 553-578 (1988).
17. T.W. Athey, "A model of the temperature rise in the head due to magnetic resonance imaging procedures, *Magnetic Resonance in Medicine* 9: 177-184 (1989).
18. D.J. Schaefer, Dosimetry and effects of MR exposure to RF and switched magnetic fields, *Annals of the New York Academy of Sciences* 649: 225-236 (1992).
19. J.P. Reilly, Peripheral nerve stimulation by induced electric currents: exposure to time-varying magnetic fields, *Med. & Biol. Eng. & Computing,* 27: 101-110 (1989).
20. J.D. Bourland, J.A. Nyenhuis, G.A. Mouchawar, L.A. Geddes, D.J. Schaefer and M.E. Riehl, Z-gradient coil eddy-current stimulation of skeletal and cardiac muscle in the dog, *in:* "Society for Magn. Res. in Medicine, 10th Annual Scientific Meeting," paper 969 (1991).
21. J.D. Bourland, J.A. Nyenhuis, G.A. Mouchawar, L.A. Geddes, D.J. Schaefer, and M.E. Riehl, Physiologic indicators or high MRI gradient-induced fields, *in:* "Society for Magn. Res. in Medicine, 10th Annual Scientific Meeting," paper 1276 (1991).
22. J.D. Bourland, J.A. Nyenhuis, G.A. Mouchawar, L.A. Geddes, D.J. Schaefer, and M.E. Riehl, Human peripheral nerve stimulation from z-gradients, in: "Society for Magn. Res. in Medicine, 9th Annual Scientific Meeting," paper 1157 (1990).
23. H. Yamagata, S. Kuhara, Y. Seo and S. Ueno, Comparison of thresholds for nerve excitation elicited by trapezoidal and sinusoidal time-varying magnetic fields in MR echo-planar imaging, *Med. & Biol. Eng. & Comput.* 29: Supplement 1991, 657 (1991).

DISCUSSION

Chairman Prof. S. Matsuoka: Thank you very much. Are there any questions or comments? Yes, please.

Prof. M. Markov: This is a question directed not only to you but to all speakers this morning. One thing that every speaker showed was that stimulation is due to induced eddy currents. You also mentioned this and that's why I ask you whether you are sure that you are just stimulating the cardiac system and not some other tissues as well. The point of the question is, how can we identify the effect produced by magnetic field stimulation? I suppose it is possible that you stimulate not just the heart muscle, but you also stimulate parts of the body which are nearer to the coil.

Dr. Nyenhuis: Sure, there was also significant skeletal muscle stimulation when the heart was stimulated. The experimental animals moved a great deal in response to the stimulating pulses. We checked to verify that this motion wasn't causing the stimulation by applying a suitable drug to paralyze the dog; that did not change the stimulation threshold. Another question might be whether the magnetic field itself influences stimulation thresholds. In a future study, we will use an MRI magnet and repeat the cardiac stimulation study to see whether there's an influence of the static field. In that regard, I've found very interesting Dr. Ueno's report that static magnetic fields have no influence on nerve conduction velocities.

Prof. S. Ueno: First of all, thank you very much for your wonderful talk. The talk includes two big jobs, one of which is cardiac stimulation. I think you're the first in the world to have succeeded in cardiac stimulation. And in the latter part of the talk, you discussed the safety problem of MRI, especially the dB/dt problem. As you kindly mentioned in your talk, yes, we are working on the same dB/dt problem with the cooperation of the Toshiba company. And you mentioned that the threshold for peripheral nerve is about 150 T/s using a single pulse. We are using a trapezoidal pulse and we obtained almost the same result, about 80 to 100 T/s. You note that cardiac stimulation requires a dB/dt of some 4000 T/s, some 30 times greater than peripheral nerve stimulation. It seems to me that with repeated pulses, the cardiac stimulation threshold would be reduced.

Dr. Nyenhuis: That's a good point, Dr. Ueno. It seems to me that there should still be an adequate cardiac safety margin for repeated pulses, but that should be looked into. The reduction of threshold, I think, would depend on how much the pulses are separated. The cardiac time constant is about 2 ms. So if pulse separation is about 2 ms, there may be a threshold reduction. On the other hand, the trapezoidal pulses produce eddy currents in the positive and negative directions, so the average charge applied to the heart by one gradient pulse is zero. This factor may tend to bring the threshold back up. In summary, it's not clear to me what influence a pulse train would have on the cardiac stimulation threshold.

Prof. Matsuoka: Thank you very much, Dr. Nyenhuis.

BIOLOGICAL EFFECTS OF EXTREMELY LOW FREQUENCY MAGNETIC FIELDS

Marko S. Markov

Department of Biophysics
Sofia University
Sofia, Bulgaria*

Recent studies have demonstrated the great significance of low-intensity, low-frequency electromagnetic fields (EMF) for living systems. It is now well-accepted that EMF may induce changes in living systems on the organism, tissue, cellular, membrane and subcellular levels. Some specific reactions and processes in different biological systems suggest that most of the observed bioeffects strongly depend on the parameters of applied electromagnetic fields.

BASIC PROBLEMS

There are at least three distinct lines of evidence tending to indicate the coupling of weak magnetic fields to living systems. These include:

-- the use of electric, magnetic or electromagnetic fields in orthopedics, physical therapy, neurology, and oncology [1-3]
-- a large number of observations showing altered cell response following exposure to electromagnetic, electric and magnetic fields [4-6]
-- concerns regarding the environmental influence of electromagnetic fields [7].

A large body of studies has suggested that different electromagnetic and magnetic fields may initiate significant bioeffects. At present, the following problems appear to be open for evaluation:

-- the search for targets of electromagnetic radiation on cellular or subcellular levels
-- examining mechanisms of EMF action on living systems
-- adaptation of living systems to applied EMF
-- evaluation of "window" effects
-- comparison of effects of ionizing and non-ionizing radiation
-- creation of standards for EMF in occupational conditions and everyday life
-- and long-lasting after-effects of electromagnetic exposure.

DOSIMETRY

All these questions are linked since the physical parameters of electromagnetic fields directly relate to the mechanism of action. It is also now well-understood that the physiological state of living systems plays an important role in their EMF sensitivity. All above-mentioned

Biomagnetic Stimulation, Edited by
S. Ueno, Plenum Press, New York, 1994

problems concern the whole range of fields under the umbrella of EMF [8]. The investigation into long-lasting effects of EMF as well as the parallel approach to ionizing and non-ionizing radiation effects have been suggested recently by Markov et al. [9]. It appears that the generalization of ionizing and non-ionizing radiation research methodology would open a new avenue in the evaluation and understanding of the nature and mechanisms of bioeffects caused by EMF action.

In spite of the large number of studies that have been done, a general approach to magnetobiological experiments is still not available. Presumably, two main reasons have to be considered: the difference in backgrounds of the scientists involved in the research, including physics, chemistry, medicine, engineering, etc.; and the variety of physical parameters of EMF.

As has been pointed out elsewhere [1], a systematic study of EMF action on any particular biological system has to consider and explain in detail the following parameters of EMF:

-- type of field
-- intensity or induction
-- gradient
-- vector
-- frequency
-- pulse shape
-- component (electric or magnetic)
-- localization
-- time of exposure

In addition, the state and properties of water inside and surrounding the biological object [10], and existence or absence of magnetic material inside the cell or in certain biological structures [11] should be considered.

The central problem appears to be identification and clarification of the interaction mechanisms between biological systems and acting electromagnetic fields. In the particular case of weak EMF effects studies, four important problems are currently discussed:

-- are electric or magnetic fields more likely to affect biological systems?
-- what are the parameters of the fields most responsible for produced bioeffects?
-- what are the biophysical and/or biochemical mechanisms at the site of interaction?
-- and, do time-varying (AC) magnetic fields affect living systems in a substantially different manner in the presence or absence of a static (DC) magnetic field?

WEAK MAGNETIC FIELD EFFECTS

The effects of weak EMF on living systems have been intensively studied and clearly established recently. Numerous experiments have demonstrated that both weak electromagnetic and magnetic fields are capable of eliciting *in vivo* and *in vitro* effects from different biological systems. These studies suggest that weak electromagnetic fields operate by modulating the cell's *in situ* control mechanisms. It was shown that exogenous weak low-frequency electric or magnetic fields can modulate certain important biochemical and physiological processes [12].

Magnetic field interactions have been reported for AC/DC magnetic field combinations with the fields oriented perpendicular or parallel [13-16] to each other. The reported data do not yet allow a clear estimation of the importance of relative AC/DC field orientation. Some authors suggest that, because of the low energy of the applied fields (compared with the chemical energy of protein binding, for example,) the observed effect should be based on strongly cooperative processes. The very important observation has been made that magnetic field energy introduced in biological systems is much less than thermal energy and many orders of magnitude less than the electric barrier of the membrane potential. These findings support the view that electromagnetic field interactions with living systems are non-linear and non-equilibratory. They are consistent with the quantum approach to processes involving long-range interactions with charged biological macromolecules.

In magnetobiological experiments, researchers need to consider the precise characteristics of not only the applied or driving field, but also those of the environmental fields, such as geomagnetic and capacitively coupled electrostatic fields. In fact, the whole range of environmental electromagnetic, electrostatic and static magnetic fields (which were found to "contaminate" the experiment site) needs a great deal of study to evaluate the fields' bioeffects and to establish standards for precisely controlled biological experiments. It should also be taken into account that each particular living system during its phylogenetic and ontogenetic development is continuously exposed to the influence of the geomagnetic field (GMF). The geomagnetic field is considered a basic ecological factor which reflects the oscillations of GMF as well as the contribution of the sun, space, and stars as additional sources of natural EMF.

For many years, the main direction of research on EMF bioeffects was concentrated on induced electric fields. In order for an EMF electrical bioeffect to be possible, the electromagnetic signal must induce sufficient voltage at the target site to be detectable above thermal noise. For many EMF signals, this may only be possible if cell-cell communication via electrical contacts (gap junctions) between cells in tissues exists [17]. This can result in EMF sensitivity several orders of magnitude higher than isolated cells and/or tissues with no, or non-functional, gap junctions. The important characteristics to assure adequate electric dosimetry are, thus, the geometry and size of the target, as well as the time rate of change - dB/dt of the magnetic field for electromagnetically induced E. Many dosimetry errors have been made by not taking both the dielectric properties of the target and its geometry into account.

MECHANISM OR MECHANISMS?

The actual biophysical mechanisms for coupling of weak magnetic fields to biological systems still remain unclear. It is evident that a physiological response can occur only after the initial physical perturbations take place. It appears that a unique mechanism for electromagnetic action upon living systems does not exist. More probable is a number of mechanisms, which act separately or in parallel depending on the specificity of the biological object and its physiological status as well as on the environmental conditions. An externally applied electromagnetic field is usually considered a vectorial perturbation in the chemical or orientational distributions of interacting molecules. The major structural changes require the presence of ions, ionized groups, and permanent or induced dipoles preferably in macromolecular structures [8].

The interaction of ions at the electrically charged interface of a cell membrane is an example of voltage-dependent processes that occur under EMF action. It is appropriate to consider the coupling of the membrane structure as a whole and of each particular element of this structure to exogenous EMF. The suggestion that membranes may serve as a major target of EMF action has been made by several authors [18-21], especially taking into account the participation of the cell membrane in signaling. In general, this coupling can significantly participate in a molecular feedback loop for cellular regulation [12].

EMF EFFECTS ON MEMBRANES

Because of their unique bioelectrochemical properties, cell membranes are assumed to be a primary receptor of any kind of electromagnetic radiation. Most results concern magnetic field effects on membrane ion transport and other membrane-mediated signal transduction processes. These studies consider the cell membrane as the primary site of interaction of applied electromagnetic fields. Some of these studies reported evidence that weak EMF triggered the binding of antibodies, hormones and enzymes to their specific binding sites on the cell membrane surface. This suggests a way by which EMF may influence signal transduction events in intact cells [22]. Based on the conception of the cell membrane as a powerful amplifier of EMF of very weak nature, Adey [23] proposed a unified picture of highly cooperative steps, which are thought to be calcium-dependent:

a) initial detection by cell surface glycoproteins as a result of electrochemical changes at binding sites;
b) transmembrane portion of protein signals these events to cell interior; and
c) coupling of the signal with the cytoskeleton.

The EMF coupling with biomembranes is usually discussed from the viewpoint of possible alterations of basic molecular components of cell membranes: lipids and proteins. While lipids are responsible for the structural integrity of the membrane, the proteins provide, via their conformational changes, structural modifications related to functional regulation. As has been pointed out, membrane proteins are thought to be a main target for EMF, more likely than the membrane lipids. It is well-accepted that both lipid and protein portions of the membrane have hydrophilic and hydrophobic components. The hydrophobic component creates a particular membrane side of the electrified interfaces at which specific binding can take place. Electromagnetic fields are considered capable of initiating changes in protein and lipid domains distribution in the membrane bilayer as well as conformational changes in lipid-protein associates.

The changes in these complexes are very tightly bound to alterations in vicinal water and double electrical layer. The fact that membranes are surrounded by hydration water is of importance for non-specific electrostatic interactions involving water dipoles and completely or partially hydrated ions. These non-specific interactions are governed by a Boltzman distribution with respect to the aqueous layer. Thus, it should be expected that water dipoles provide the "first" layer of charge interactions with the membrane surface. Water dipoles followed by a rigid layer of hydrated ions form a strongly equilibratory structure which can be perturbed to a negligible degree by an injection of a small amount of dipoles or charges.

However, a second type of interaction at the membrane surface can also occur, involving potential dependent specific binding. Under these conditions, an ion or dipole can compete with water dipoles and hydrated ions for their specific membrane sites. This event involves the subsequent steps of dehydration, displacement and binding. Thus, a modification of membrane structure occurs and is reflected in enzyme activity and membrane transport.

It appears that membrane transport should be considered in three distinct directions: the transport of materials, of energy, and of information.

The first two directions are mainly discussed in terms of membrane and cellular biophysics, while the third still remains insufficiently studied, due to the fact that the information aspect of biological processes had not been studied until recently. It appears that in some cases, instead of discussing "transport," it is better to discuss "utilization" of information that reaches the membrane. The informational approach could be of great importance, especially when weak EMF interactions with biosystems are considered. It seems quite probable that under certain conditions, the presence or the absence of the magnetic field is more important for living systems than the amplitude of the applied field.

In this respect, the alteration of both active and passive electrical properties of membranes can assure important information concerning effects of magnetic fields at the membrane level. A pulse impedance method, proposed by Pliquett [24] provides the possibility for a very fast screening of a variety of solutions. The principle of the method consists of applying a rectangular electrical pulse to the investigated sample. The modified signal as a response of the object to the applied physical factor may be visualized on the screen of an oscilloscope (Figure 1). The most informative parameter was found to be the value of Δ h. Blood prepared for transfusion was exposed to a static magnetic field of 45 mT, and a specific dynamic of change of Δ h values was observed (Figure 2). This appears to suggest that magnetic fields of 45 mT may prevent hemolysis of red blood cells and thus prolong the period of use of blood for transfusion [25].

The same method was used for evaluation of the action of membrane charge modificators. It was found that a membrane is capable of binding the positively charged protamine, while the effects of negatively charged heparin are manifested at the level of the membrane's

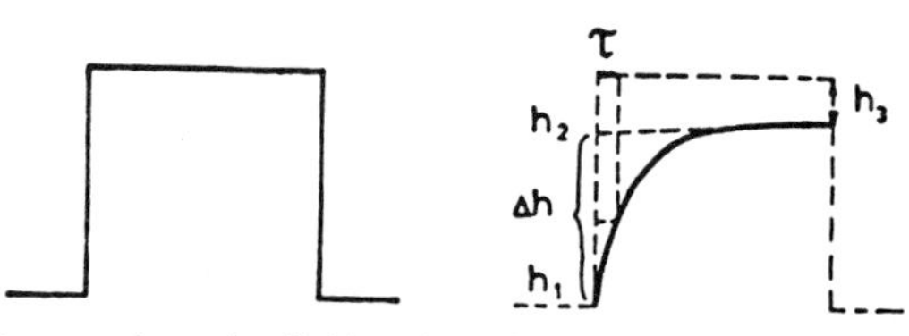

Figure 1. Rectangular pulse (left) and modified response of cell suspension (right).

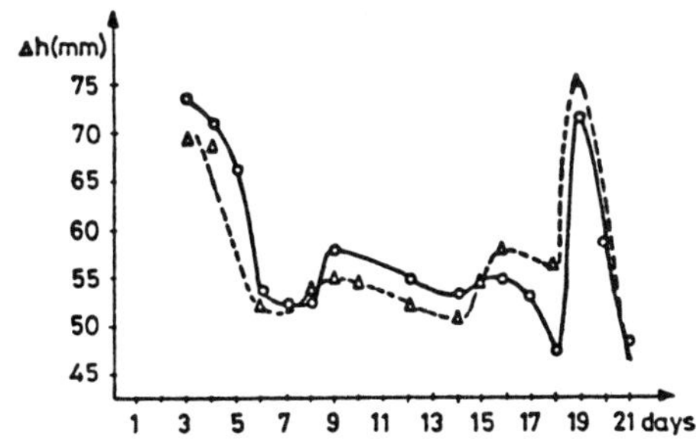

Figure 2. Change in the parameter Δh during blood conservation. The solid line is the control, the dashed line is the magnetic field-exposed blood.

double electrical layer. The magnetic field modifies the binding of both protamine and heparin, and this suggests that magnetic fields of 10 to 50 mT may be used for certain therapeutical applications. It is common practice to apply magnetic fields of this field strength for treatment of a number of diseases. However, this raises the problem of clear and exact diagnosis and a choice of appropriate method of therapy with magnetic fields for a given patient and disease.

WINDOW EFFECTS

In the last decade, the concept of "biological windows" began to be discussed and studied. The sensitivity of biological systems to weak magnetic fields has been described elsewhere [6, 8, 10, 27-31], mainly in respect to the dependence of bioeffects on the amplitude and frequency of applied signals, as well as on duration of exposure.

In a series of experiments designed to study a large range of inductions (up to 100 mT) of static magnetic fields applied to biosystems with different levels of organization (animal, microorganisms, plants), a specific dependence of the observed bioeffects upon the amplitude of applied field was found at induction of 45 mT [32,33]. These results clearly indicate the existence of specific "permitted" levels which biosystems could attain under the action of a static magnetic field with a certain amplitude. When the field amplitude is equal to those necessary for the transition, the system achieves a "stationary" state at which it can remain for a certain period of time. Any other magnetic fields inject energy into the system, which stimulates a transition to a state different from the stationary one. Under these conditions the system intensively exchanges energy with the environment, and because of that, the observed effects are smaller and they quickly disappear at these "non-permitted" states [10].

In many respects, this idea is similar to suggested "window" effects at very weak electromagnetic fields [4, 5, 27]. Several recently published studies show that imposed extremely low-frequency electric or magnetic fields can modulate certain important biochemical and physiological processes [18, 35-37]. Most of the evidence for frequency and amplitude windows is connected with the state and movement of calcium ions [38].

"Window" or "resonance" effects are present at RF field amplitudes modulated with low-frequency magnetic fields, at low-frequency electric fields, at low-frequency EMF, as well as at static magnetic fields. This problem is closely related to adaptation of living systems to any change in electromagnetic background, mainly because of the already mentioned differences in the sensitivity of biological systems toward specific amplitudes and frequencies of the applied fields. Note that many of the living systems studied react more to very weak magnetic fields than they do to stronger fields.

A number of experiments performed by us examined a large interval of induction of static magnetic fields applied to various biological systems (microorganisms, plants, animals) [32, 33].

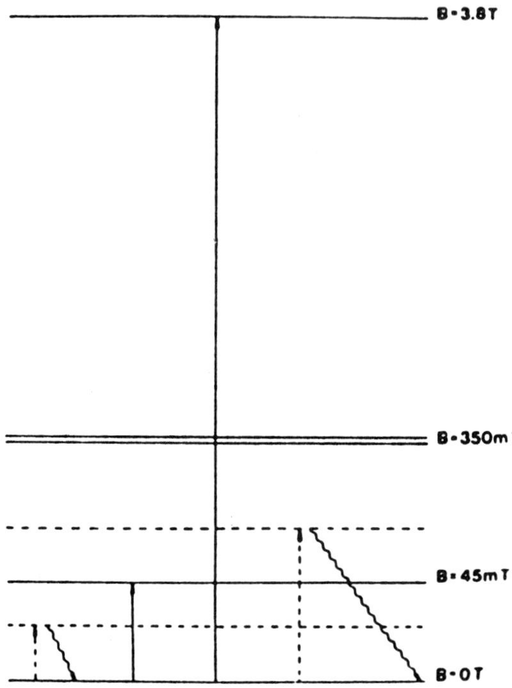

Figure 3. "Resonance" model for amplitude windows.

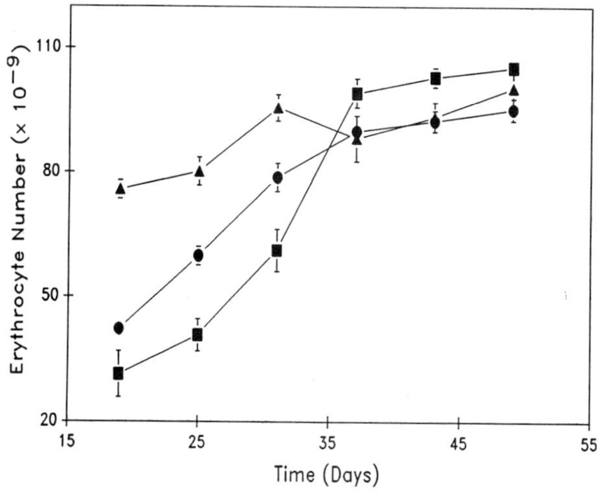

Figure 4. Changes in erythrocyte counts of rats. The triangles represent the control group, the squares represent the group of direct magnetic field action, and the closed circles represent the group of indirect action.

In these experiments, a specific dependence of the observed bioeffects on the amplitude of magnetic field was found, with an optimum at a magnetic induction of 45 mT (Figure 3).

It was shown elsewhere that the exposure of new-born rats to static magnetic fields in the interval of 10 to 100 mT clearly follow the "resonance" hypothesis. Results presented in Figures 4 and 5 suggest a very important conclusion concerning adaptational properties of living systems as demonstrated on the organism level. In these experiments, three groups of Wistar rats were the subject of investigation.

The first group contained animals that had been individually exposed for 30 minutes to a static magnetic field of 45 mT. The second group of animals, the group of indirect action, was given a daily ration of water that had been preliminarily exposed to the same magnetic field. The third group was a control. Each group contained at least six animals. During the 48-day period of investigation, several blood parameters were controlled, including the number of erythrocytes and leukocytes, the quantity of hemoglobin, and sedimentation rate.

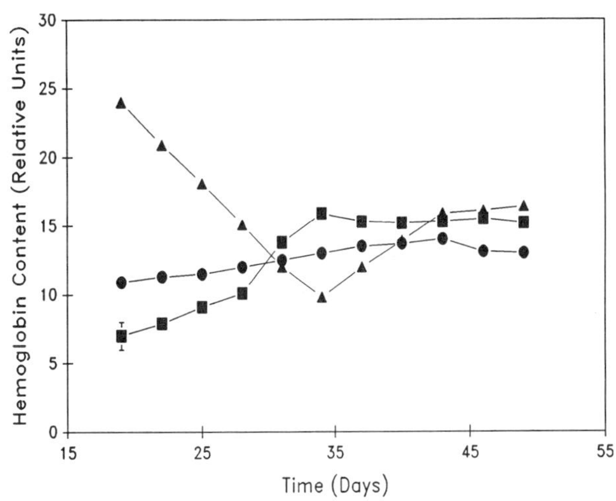

Figure 5. Changes in the hemoglobin content of rat blood. The triangles represent the control group, the squares the group of direct magnetic field exposure, and the circles the group of indirect action.

Table 1. Average values and standard deviation of trace elements in blood.[1]

Element	Control (N = 15)	Direct Magnetic Field action (N = 18)	Indirect Magnetic Field action (N = 18)
Fe	(60.6 ± 5.1) mg%	(38.8 ± 1.7) mg%	(68.5 ± 4.4) mg%
Zn	(1.67 ± 0.06) mg%	ND	ND
Si	(2.85 ± 0.18) mg%	(1.13 ± 0.21) mg%	(5.37 ± 0.27) mg%
Pb	(329.2 ± 27.9) µg%	(283.3 ± 40.0) µg%	(378.6 ± 21.6) µg%
Al	(85.4 ± 11.8) µg%	(51.9 ± 6.9) µg%	(55.0 ± 7.7) µg%
Cu	(784.8 ± 55.6) µg%	(102.1 ± 10.6) µg%	(440.2 ± 32.2) µg%

Table 1. N is the number of animals in the group, and ND stands for "not detected".

All these parameters behaved identically. There was a large deviation in the first 25 days, followed by a gradual equalization of the parameters. This was considered evidence for the adaptation of growing animals to everyday "magnetic stress." After the exposure period was over, animals were sacrificed and the emission spectrum analysis of the amount of several microelements was performed. It was observed that a serious redistribution of microelements in blood, especially Zn and Cu, took place as a result of exposure to static magnetic field (Table 1). Moreover, this redistribution concerned different organs. It appears that the observed decrease of Cu and Zn content in blood is accompanied by a parallel increase of the amount of these microelements in other organs, especially the spleen and liver. For instance, the content of Zn in the spleen was found to increase by 711 percent.

In addition, the long-lasting effects of electromagnetic fields should be considered in connection with the short-term adaptation. Data about long-lasting effects are completely insufficient and unsystematic. Only serious collection and consideration of data may shed light on this problem. This is especially important for human organisms who are professionally engaged with electrical equipment and technologies in which strong magnetic fields are present.

MAGNETIC FIELD EFFECTS ON CELL-FREE SYSTEMS

Certain attempts to estimate magnetic field effects at the subcellular level were also carried out. Magnetic field interactions have been reported for AC/DC magnetic fields oriented perpendicular or parallel to each other. Again, the suggestion of strong cooperativity of bioeffects, due to the low energy of applied signals, has to be considered. The search for experimental evidence and theoretical mechanisms for the coupling of magnetic fields and phosphorylation of some specific enzymes, especially those participating in muscle activity, seems to be a quite promising approach. Three groups of intracellular enzymes were found to respond to membrane signals, which appeared when electromagnetic fields were applied:

- -- membrane-bound adenylate involved in activation of protein kinase through conver sion of ATP to cAMP as explored in bone cells [39, 40];
- -- cAMP-independent protein kinase that performs messenger functions [41];
- -- and ornithine decarboxylase, essential for cell growth and involved in synthesis of polyamines which are essential for DNA formation [42].

Very recently, the effects of extremely weak (environmental range) sinusoidal (AC), and DC magnetic fields on myosin light change phosphorylation in a cell-free system were studied by us [37, 43]. This enzyme system is Ca^{2+}-calmodulin dependent, providing a well-established ion-binding mechanism. The exposure system applied DC (0-200 μT, vertically and/or horizontally controlled) and/or sinusoidal AC (16 Hz, 20.9 μT) magnetic fields. The results appear to suggest that all of the AC, DC, as well as combinations of AC/DC magnetic fields significantly (30 to 100 percent) influenced myosin phosphorylation. The most surprising and important finding was that variations of the DC magnetic field (in the absence of AC components) not only were sufficient to alter the rate of phosphorylation, but also gave the maximum effect.

Experiments with μ-metal shielding, which decreased the values of ambient DC and all AC magnetic fields to plus or minus 0.1 μT, also resulted in a decrease of phosphorylation. As seen in Figure 6, the AC field alone affects myosin phosphorylation by more than 40 percent, and the simultaneous application of AC/DC results in more than 50 percent inhibition of phosphorylation. A very important result was that the DC magnetic field alone modifies phosphorylation more strongly than either AC alone or the AC/DC combination, attaining 39 percent of the mean control (ambient) value.

In these, as well as in most of the following experiments, the horizontal component of the ambient DC magnetic field was actively compensated to zero. These surprising results suggested the design of a specific experiment that allowed the evaluation of the effect of the vertical DC magnetic field (0-200 μT), on myosin phosphorylation in the absence of any horizontal DC and any applied AC fields. When the DC magnetic field was reduced to 0-50 μT (below ambient), the rate of phosphorylation decreased to 35 to 50 percent of control values. In contrast, phosphorylation increased to approximately 125 percent compared to control, in the range of 110 to 140 μ (where the vertical DC is about twice the ambient).

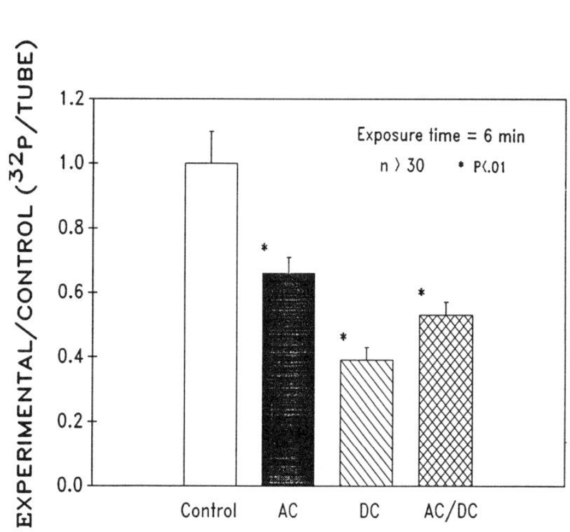

Figure 6. The effects of extremely weak AC and DC magnetic fields on myosin phosphorylation. Vertical DC was 20.9 μT, horizontal DC was compensated to zero in all experiments, and AC was 20.9 μT, at 16 Hz sinusoidal; for the control, both components of ambient magnetic field were present.

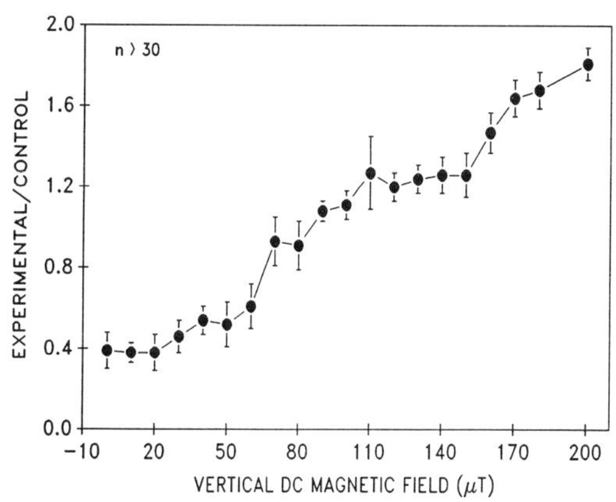

Figure 7. Vertical DC magnetic field effects on myosin phosphorylation. The horizontal component of the ambient magnetic field was compensated to zero.

When fields higher than 150 μT were applied, the rate of phosphorylation monotonically increased to more than 180 percent. These results are shown in Figure 7. Shielding of all AC and DC magnetic fields leads to a significant inhibition of the rate of phosphorylation. However, the application of a 200 μT vertical DC magnetic field, within the μ-metal shield, increased phosphorylation efficiency for any exposure (2 to 15 minutes). From these results, it could be concluded that initiation of the changes in myosin phosphorylation rate occurred at an early stage of the reaction exposure time, which then acted as a trigger for those phosphorylation stages that are actually measured in this study.

The results presented in this paper, as well as the effects already reported by us [37, 43], clearly demonstrate that weak magnetic fields are able to modulate this biochemical reaction in a cell-free system. However, the theoretical interpretation of the observed effects is not yet clearly established. Several theories have been proposed to explain the effects of such weak magnetic fields. Nevertheless, the actual biophysical pathways for coupling of weak magnetic fields to biological systems still remain unclear. It is evident that a physiological response can occur only after the initial physical perturbations take place. The above presented results do not support either ion cyclotron resonance [44] or parametric resonance theories [35].

As has been pointed out [45], there are at least five possible targets for magnetic field action in this cell-free system: the calcium ion, the calmodulin molecule, the substrate or myosin light chain, the enzyme or myosin light chain kinase, and the ATP molecule. At present, there is no clear evidence as to which of these targets is responsible for the observed effects. From general considerations, the most promising target appears to be calmodulin, and especially its Ca^{2+} binding sites. It appears quite possible that the probability of Ca^{2+} binding may be altered by the applied DC fields. It is also possible that the structure or energetics of the hydrated Ca^{2+} complex may be modified, which in turn may alter its binding to calmodulin. If this is found to be true, serious reconsideration of participation of Ca^{2+} in a large number of magnetic field bioeffects should be reconsidered.

In summary, the reported set of experiments has quite clearly shown that even very small alterations in the ambient DC magnetic fields are sufficient to have a profound effect on a large variety of biological systems. The results strongly suggest that living systems can have exquisite sensitivity to weak electromagnetic and magnetic fields. This paper is clearly in favor of the viewpoint that many reported bioeffects are purely magnetic field effects, rather than having an electrical base.

---

*At present the author is associated with the Bioelectrochemistry Laboratory, Department of Orthopedics, Mount Sinai Medical Center, 1 Levy Place, New York, NY 10029-1188.0

REFERENCES

1. A.T. Barker, R.A. Dixon, W.J.W. Sharrard, and M.L. Sutcliffe, Pulsed magnetic field therapy for tibial non-union, *Lancet* 994 (1984).
2. C.T. Brighton, Advanced clinical applications of EMF effects: Bone and cartilage, *in:* "Electromagnetics in Medicine and Biology," C.T. Brighton and S.R. Pollack, eds., San Francisco Press Inc., 293-308 (1991).
3. C.A.L. Bassett, Physical and biological principles affecting weak ELF electromagnetic bioresponses, *in:* "Electromagnetics in Medicine and Biology," C.T. Brighton and S.R. Pollack, eds., San Francisco Press Inc., 1-13 (1991).
4. S.M. Bawin, L.K. Kaczmarek and W.R. Adey, Effects of modulated VHF fields on the central nervous system, *Ann NY Acad Sci* 247: 74-91 (1975).
5. C.F. Blackman, S.G. Benane, D.E. House and W.T. Joines, Effects of ELF (1-120 Hz) and modulated (50 Hz) RF fields on the efflux of calcium ions from brain tissue in vitro, *Bioelectromagnetics* 6: 1-11 (1985).
6. M.S. Markov, Electromagnetic field influence on membranes, *in:* "Interfacial Phenomena in Biological Systems," M. Bender, ed., Marcel Dekker, 171-192 (1991).
7. M.S. Markov, Electromagnetic fields - new ecological factor, *in:* "Electromagnetic Fields and Biomembranes," M.Markov and M. Blank, eds., Plenum Press, New York, 135-140 (1988).
8. M.S. Markov, Effect of radiation on biological systems, *in:* "Charge and Field Effects in Biosystems," M.J. Allen, S.F. Cleary and F. Howkridge, eds., Plenum Press, New York, 241-250 (1989).
9. M.S. Markov, S.P. Ivanov, B.P. Galutzov, L.L. Traikov and M.A. Kuzmanova, Biophysical test for estimating the effects of ionizing and non-ionizing radiation, *Medical and Biological Engineering and Computing,* (in press).
10. M.S. Markov, Influence of constant magnetic field on biological systems, *in:* "Charge and Field Effects in Biosystems," M.J. Allen and P.N.R. Usherwood, eds., Abacus Press, Kent, England, 319-329 (1984).
11. J.L. Kirshvink and A. Kobayashi-Kirshvink, Magnetite biomineralization in human tissue: a solution to the thermal noise problem in ELF bioeffects, *in:* "Proceedings of First World Congress on Electricity and Magnetism," San Francisco Press Inc. (in press).
12. W.R. Adey and A.S. Sheppard, Cell structure ionic phenomena in transmembrane signaling to intracellular enzyme systems, *in:* "Mechanistic Approaches to Interactions of Electric and Magnetic Fields with Living Systems," M. Blank and E. Findl, eds., Plenum Press, New York 365-387 (1987).
13. C.F. Blackman, S.G. Benane, J.R. Rabinowitz, D.E. House, and W.T. Joines, A role for the magnetic field in the radiation-induced efflux of Ca-ions from brain tissue in vitro, *Bioelectromagnetics* 6: 327-333 (1985).
14. C.F. Blackman, S.G. Benane, D.E. House and D.J. Elliot, Importance of alignment between local DC magnetic field and an oscillating magnetic field in response of brain tissue *in vivo* and *in vitro*, *Bioelectromagnetics* 11:159-167 (1990).

15. M.F. Yost and R.P. Liburdy, Time-varying and static magnetic fields act in combination to alter calcium signal transduction in the lymphocyte, *FEBS Letters* 296: 117-122 (1992).
16. A.R. Liboff, R.J. Rozek, M.L. Shermann, B.R. McLeod and S.D. Smith, Ca^{45} cyclotron resonance in human lymphocytes, *J. Bioelectricity* 6:13-22 (1987).
17. A.A. Pilla, P.R. Nassar and J.J. Kaufman, The sensitivity of cells and tissues to weak electromagnetic fields, *in:* "Charge and Field Effects in Biosystems III," M.J. Allen, S.F. Cleary, A.E. Sowers and D.D. Shillady, eds., Birkhauser, Boston, 231-241 (1992).
18. M. Markov and M. Blank, eds., "Electromagnetic Fields and Biomembranes," Plenum Press, New York (1988).
19. A.A. Pilla, Electrochemical information transfer at cell surfaces and junctions -- application to study and manipulation of cell regulation, *in:* "Bioelectrochemistry," H.Keyzer and F. Gutmann, eds., Plenum Press, New York, 333-395 (1980).
20. A. Chiabrera, M. Hinsenkamp, A.A. Pilla, J.T. Ryaby, D. Ponta, F. Beltrame, M. Grattarola and C. Nicolini, Cytofluorometry of electromagnetically controlled cell differentiation, *J. Histochem and Cytochem* 27: 375 (1979).
21. G. Collaccico and A.A. Pilla, Electromagnetic modulation of biological processes: Chemical, physical and biological correlations in the Ca uptake by embryonal chick tibia *in vitro*, *Bioelectrochemistry and Bioenergetics* 10: 119-131 (1983).
22. E. Saalman, B. Nordenm, L. Arvidsson, Y. Hamnerius, P. Hojevik, K.E. Connell and T. Kuruscev, Effect of 2.45 GHz microwave radiation on permeability of unilamellar liposomes to 5 (6)-carboxy-flourescein, *BBA* 1064: 124-130 (1991).
23. W.R. Adey, The sequence and energetics of cell membrane transductive coupling to intracellular enzyme systems, *Bioelectrochemistry and Bioenergetics* 15: 447-456 (1986).
24. F. Pliquett, Passiv elektrische Untersuchungen biologische Objekte, Wiss. Z. Universitat Leipzig, Mathem. Nat. wiss. Reihe 28: 118-127 (1979).
25. M.S. Markov, F. Pliquett, M.R. Kantcheva and D.L. Abadgieva, Pulse impedance method for investigating electrical properties of cells, *International Agrophysics* 1: 345-355 (1985).
26. M.S. Markov and N.G. Todorov, Electromagnetic field stimulation of some physiological properties, *Studia Biophysica* 99: 151-156 (1984).
27. W.R. Adey, Evidence for cooperative mechanisms in the susceptibility of cerebral tissue to environmental and intrinsic electric field, *in:* "Functional Linkage in Biomolecular Systems," F. Schmitt, D.M. Schneider and D.M. Crithers, eds., Raven Press, New York, 325-342 (1975).
28. W.R. Adey, Models of cerebral cells as substrates for informational storage, *Biosystems* 8: 163-176 (1977).
29. W.R. Adey, The sequence and energetics of cell membrane transductive coupling to intracellular enzyme systems, *Bioelectrochemistry and Bioenergetics* 15: 447-456 (1986).
30. W.R. Adey, The extracellular space and energetic hierarchies in electrochemical signalling between cells, *in:* "Charge and Field Effects in Biosystems," M.J. Allen, S.F. Cleary and D. Howkridge, eds., Plenum Press 263-290 (1989).
31. M.S. Markov, Informational character of magnetic field action on biological systems, *in:* "Biophysical and Biochemical Information Transfer in Recognition," K. Jensen and Yu.Vassileva, eds., Plenum Press, New York 496-500 (1979).
32. M.S. Markov, Biological mechanisms of the magnetic field action, *IEEE Transaction of magnetics - v.Mag-17,* N.5, 2334-2337 (1981).
33. M.S. Markov, Direct and indirect action of constant magnetic field, "Proceedings of 6th International Conference on Magnet Technology," Bratislava, 384-389 (1978).
34. A.R. Liboff, B.R. McLeod and S. Smith, Resonance transport in membranes, *in:* "Electromagnetics in Biology and Medicine," C.T. Brighton and S.R. Pollack, eds., San Francisco Press Inc., 67-77 (1991).
35. V.V. Lednev, Possible mechanism for the influence of weak magnetic fields on biological systems, *Bioelectromagnetics* 12: 71-74 (1991).
36. L.A. Shuvalova, M.V. Ostrovskaja, E.A. Sosunov and V.V. Lednev, Weak magnetic field influence on the speed of calmodulin dependent phosphorylation of myosin in solution, *Dokladi Academy of Sciences USSR* 217: 227-231, (1991), in Russian.
37. M.S. Markov, J.T. Ryaby, J.J. Kaufman and A.A. Pilla, Extremely weak AC and DC magnetic fields significantly affect myosin phosphorylation, *in:* "Charge and Field Effects in Biosystems 3," M.J. Allen, S.F. Cleary, A.E. Sowers and D.D. Shillady, eds., Burkhauser, Boston 225-230 (1992).
38. C.V. Byus, R.L. Lundak, R.M. Fletcher and W.R. Adey, Alterations in protein kinase activity following exposure of cultured lymphocytes to modulated microwave fields, *Bioelectromagnetics* 5: 34 (1984).
39. R.A. Luben, C.D. Cain, M.C.Y. Chen, D.M. Rosen and W.R. Adey, Effects of electromagnetic stimuli on bone and bone cells in vitro; inhibition of responses to parathyroid hormone by low energy, low frequency fields, *Proc. Nat'l. Acad. Sci. USA* 79: 4180-4184 (1982).
40. R.A. Luben, C. Cain, Use of bone cell hormone responses to investigate bioelectromagnetic effects on membranes in vitro, *in:* "Nonlinear Electrodynamics in Biological Systems, W.R. Adey and A.F. Lawrence, eds., Plenum Press, New York, 23-33 (1984).

41. C.V. Byus, R.L. Lundak, R.M. Fletcher and W.R. Adey, Alterations in protein kinase activity following exposure of cultured lymphocytes to modulated microwave fields, *Bioelectromagnetics* 5: 34-51 (1984).
42. W.R. Adey, Cell membranes: The electrochemical environment and cancer promotion, *Neurochemical Research* 13: 671-677 (1988).
43. M.S. Markov, J.T. Ryaby, S. Wang and A.A. Pilla, Modulation of myosin phosphorylation rates by weak (near ambient) DC magnetic fields, *in:* "Proceedings of 1992 IEEE 18th Annual Northeast Bioengineering Conference," Kingston RI, p.63-64 (March 1992).
44. A.R. Liboff, Cyclotron resonance in membrane transport, *in:* "Interactions Between Electromagnetic Fields and Cells," A. Chiabrera, C. Nicolini and H.P. Schwan, eds., Plenum Press, New York, 281 (1985).
45. M.S. Markov, S. Wang and A.A. Pilla, Effects of weak low frequency sinusoidal and DC magnetic fields on myosin phosphorylation in a cell-free preparation, *Bioelectrochemistry and Bioenergetics,* (in press).

DISCUSSION

Chairman Prof. I.L. Freeston: Thank you very much, Professor Markov. Are there any questions or comments?

Prof. P.Å. Öberg: Thank you for your interesting presentation. I would like to hear your opinion regarding the debate going on about magnetic fields around computer terminals and personal computers. Do we have bioeffects from computers or not?

Prof. Markov: Thank you for your question. It is really out of my personal experience, but I had the chance to attend two meetings in which this problem was discussed.

The first meeting was in 1987 in Pushchino (in the USSR,) where for the first time I heard that many computer operators claimed that after a couple of hours of work, they felt unable to concentrate on work, became nervous, and some of them got headaches. The authors of this presentation said that the main reason was not the flickering of the screen, but the existence of weak AC magnetic fields.

Two weeks ago, at the annual meeting of the Bioelectromagnetics Society in Salt Lake City in the USA, this question was again discussed. Summarizing the discussion at both meetings, it appears that today nobody doubts the existence of weak magnetic fields, which probably create some complications to computer operators. At present, nobody knows what kind of disturbances may be expected, because no systematic studies have been performed.

Nevertheless, in the case of adults who work with computers, EMF effects may be considered a factor of professional and occupational risk. The problem becomes much more serious with respect to children who spend hours playing computer games in addition to work they perform in school as part of their educational program. To my mind, the effects of EMF radiation on children has become a problem of great importance.

Prof. S. Ueno: Thank you very much for sharing your interesting results. Especially in the last four slides, one can see that your data are completely different from the results of Dr. Liboff's, Dr. Adey's and Dr. Blackman's groups. The cyclotron resonance model proposed by Liboff is quite different from your data, because they applied both DC and AC (16 Hz) weak magnetic fields. In your case, without any AC field you obtained these extremely interesting results. Could you comment on that?

Prof. Markov: I understand the reason for this question. If you look at my transparencies showing the combined AC and DC magnetic field, you should see a coincidence with the reported data and, in this respect, they confirmed the theory of Liboff, Chiabrera, and Lednev.

But results we got with only DC fields are completely inconsistent with the models. Note that all resonance models required some fixed amplitudes and frequencies for occurrence of the resonance. On the other hand, the quantum approach recently proposed by Lednev and further developed by Chiabrera looked at the effects of splitting the Zeeman levels. A weak

DC magnetic field is necessary for this splitting. For this reason, you could expect that DC alone may initiate some biochemical and biophysical alteration.

Prof. Ueno: One more, very simple question. You obtained very interesting results at 20.9 μT magnetic field, and you didn't get the same results using 200 μT field. So, I wonder what kind of mechanism you would suggest for these results? And also, what is the contribution of surrounding sources of magnetic fields?

Prof. Markov: I didn't have enough time to explain the exact protocol of my experiments. You should take into account that in all experiments, I cancelled the horizontal component of the earth's magnetic field and thus my conditions are different from the ambient, usual conditions.

The reported data clearly show that a simple biological system such as a single cell or subcellular structure is very sensitive to small changes in the electromagnetic background. Perhaps in the case of a whole organism, as I showed in my previous experiments, some systems are engaged in adaptation and compensation of the effects occurring from any new physical or chemical factor. This could be the reason that humans do not suffer when electromagnetic conditions change. But, for how long this compensation may work, I don't know. Possibly, if you stay long enough in changed conditions, you will feel some discomfort, but for how long? Who knows?

Prof. Freeston: Are there any other questions or comments? Then, thank you very much, Professor Markov.

MAGNETIC STIMULATION OF NERVE TISSUE

P. Åke Öberg

Department of Biomedical Engineering
University of Linköping
S-58185 Linköping
Sweden

Mr. Chairman, Ladies and Gentlemen; let me first thank Professor Ueno for inviting me to this interesting seminar.

Today I am going to talk about three different project types, all concerning magnetic stimulation of nerve tissue. The first project concerns stimulation of peripheral nerves with a switched magnetic field. In about 1970, I was involved in a project that was studying the dynamics of arterial blood pressure regulation. We wanted to study the dynamic response of the arterial blood pressure regulatory system after a disturbance that was generated by electrical stimulation of the carotid sinus nerve, an important feedback link in the blood pressure control system.

We used long-term, chronic dogs, meaning dogs that had a stimulation device permanently implanted, and whose carotid artery on one side was placed in a skin tube to permit repeated puncture for accurate systemic blood pressure measurements. We then put platinum electrodes on the carotid sinus nerve and we stimulated non-invasively through the skin via inductive coils. We very soon found that the stimulation effect decreased with time, which had to do with formation of fibrous tissue which isolated the nerve bundle from electrodes electrically.

So, we were interested in developing a stimulation principle that worked better in long-term chronic dogs than the electric one. Switched magnetic fields were one possibility to investigate. In physiology departments, the frog nerve-muscle preparation has been used as an excitation model since the time of Galvani. So we used nerve-muscle preparations from frogs to compare the effects of two kinds of stimulation: regular electrical stimulation with two magnetic stimulation methods.

Figure 1. The two types of magnetic stimulation were (1) bursts driving a ferrite core with an airgap, and (2) using decaying stimulation with a ringing circuit to produce a magnetic field. With this set-up we could stimulate and record the isometric contraction force of the muscle, the twitches, or the tetanus type of responses. We could also, in some experiments, record the electrical activity in the nerve. In this way we optimised the stimulation parameters in this very simple model set-up.

A comparison between electric and magnetic stimulation of this preparation, using the ringing type of stimulation, is shown in Figure 2.

Figure 2. The isometric muscle force was measured with a special transducer. The two stimulation types develop roughly the same maximum force in the muscle, but there is a slight difference in the time course of the response. The force developed with electrical stimulation has a shorter duration than that developed with the magnetic type.

Biomagnetic Stimulation, Edited by
S. Ueno, Plenum Press, New York, 1994

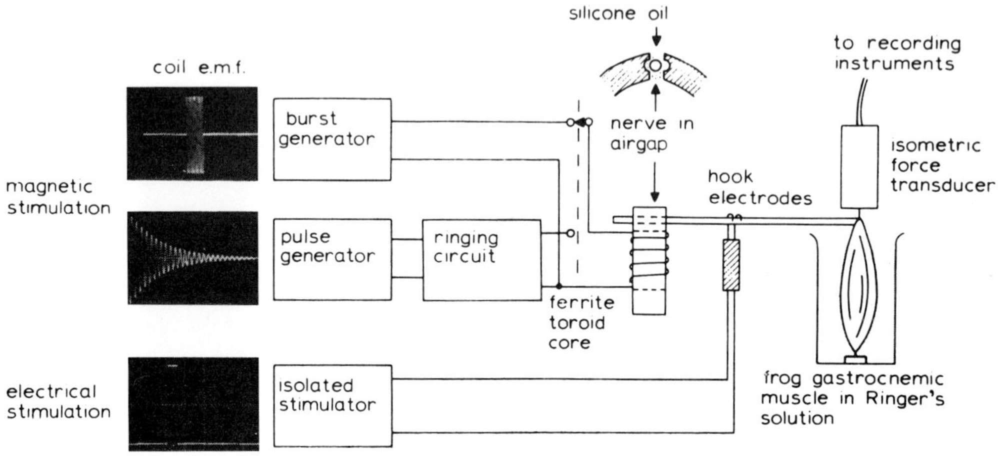

Figure 1. Experimental set-up.

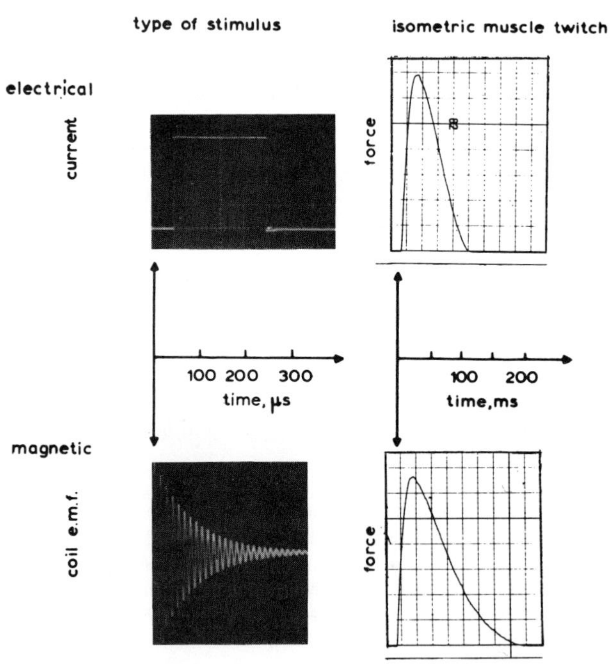

Figure 2. Comparison between the processes of contraction with electric and magnetic stimulation (decaying types).

Figure 3. This figure demonstrates a comparison between magnetic and electric stimulation with different duration. The left part of the figure shows the muscle contraction force as a function of time when the corresponding nerve is stimulated during 1, 5, 5, 50 and 425 milliseconds. As can be seen, the force developed after magnetic stimulation is increasing with the stimulation duration and resulting in a tetanic contraction at long stimulation periods. During electrical stimulation via the electrodes on the nerve, the muscle responds with a twitch with a slightly varying amplitude. At longer stimulation durations we can easily see that the muscle responds both at the onset of stimulation and at the end of the stimulation period. So there is considerable difference in response between the two ways of stimulating the muscle.

We also looked at the effect of the frequency within each burst. When varying the frequency from 1 kilohertz up to 40 kilohertz, we got an unchanged response up to around 10 kilohertz as shown in Figure 4.

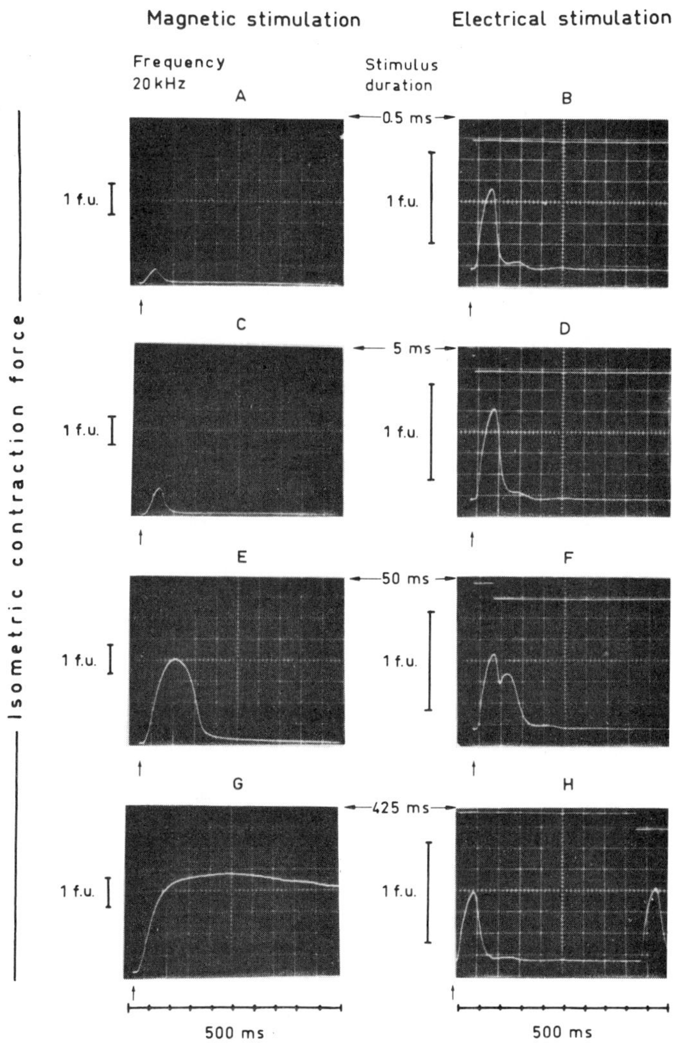

Figure 3. Comparison between the isometric processes of contraction in the gastrocnemius muscle in magnetic stimulation (A, C, E, G) and electric stimulation (B, D, F, H) of the ischiadicus nerve. The periods of stimulation were 0.5, 5, 50, and 424 ms. 1 f.u. = 1 force unit.

Figure 4. Above this frequency, the muscle force was dropping, probably indicating a less efficient stimulation of the sciatic nerve. The burst duration and its relation to the magnetic field's ability to elicit a response was studied in the same type of experimental set-up.

Figure 5. This figure shows the result. The curve is similar to the classic strength-duration curve. When we increased the burst duration, less magnetic flux density in the airgap was required to elicit a contraction response.

The magnetic stimulator was made with a ringing parallel resonance circuit the Q-value (damping) of which can be varied electrically. By means of this circuit we could grade the magnitude of the response of the muscle as can be seen in Figure 6.

Figure 6. Three different damping conditions of the ringing circuit resulted in graded variations in the muscle twitch force.

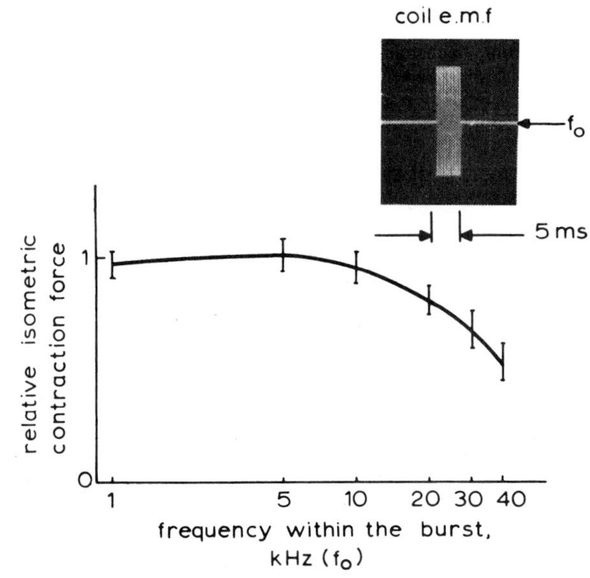

Figure 4. Contraction-force maximum as a function of the frequency within each burst. The duration of the burst was 5 ms at all frequencies.

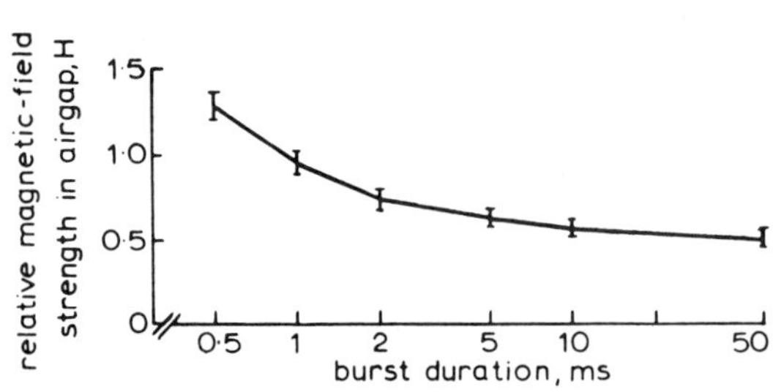

Figure 5. The strength/duration relationship with magnetic stimulation of the nerve-muscle preparation. The frequency within each burst was 20 kHz.

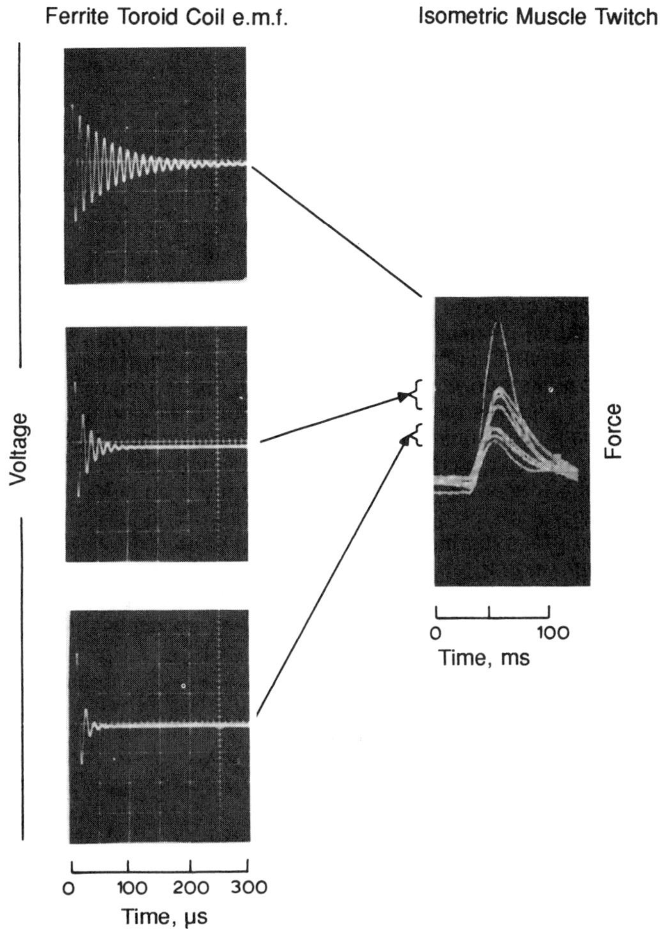

Figure 6. Process of contraction of the muscle with different degrees of electrical attenuation of the ringing circuit.

HUMAN RETINA STUDIES

The magnetic stimulation of nerve-muscle preparations generated an interest in how magnetic fields affect human beings in a more global respect. At that time (1973) one of the few reproducible effects of the interaction between a low-frequency magnetic field and a biological system was the magnetophosphene, a phenomenon that has been known since 1896 when D'Arsonvale described it for the first time. Magnetophosphenes are a kind of pale, flickering optic phenomena that you can see if you expose your head to low-frequency magnetic fields. Many groups have been studying this phenomenon through the years, but we were interested in more quantitative data, for instance threshold data for the frequency interval 0-50 Hz and magnitude data in the range 0-100 mT.

Figure 7. Our set-up for the study of magnetophosphenes is illustrated in Figure 7, which shows the electromagnet positioned over the temples, a current amplifier producing a current through the magnetic coil and a low-frequency wave generator. We very soon discovered that the background light in the laboratory affected our results strongly, so we created a homogenous background light level by shining light into a large semi-sphere with a white interior. In this set-up we could easily generate background light of various spectral compositions and wavelengths. This is how we studied the magnetophosphenes and the threshold to elicit magnetophosphenes as a function of the magnetic stimulation frequency in the frequency range 1-50 Hz.

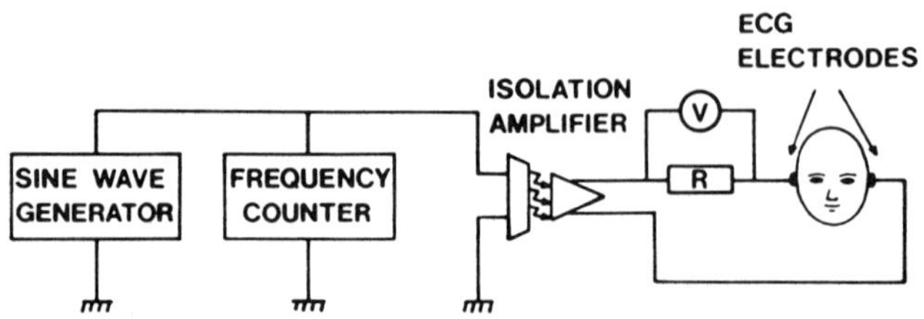

Figure 7. Experimental set-up for generation of electrophosphenes.

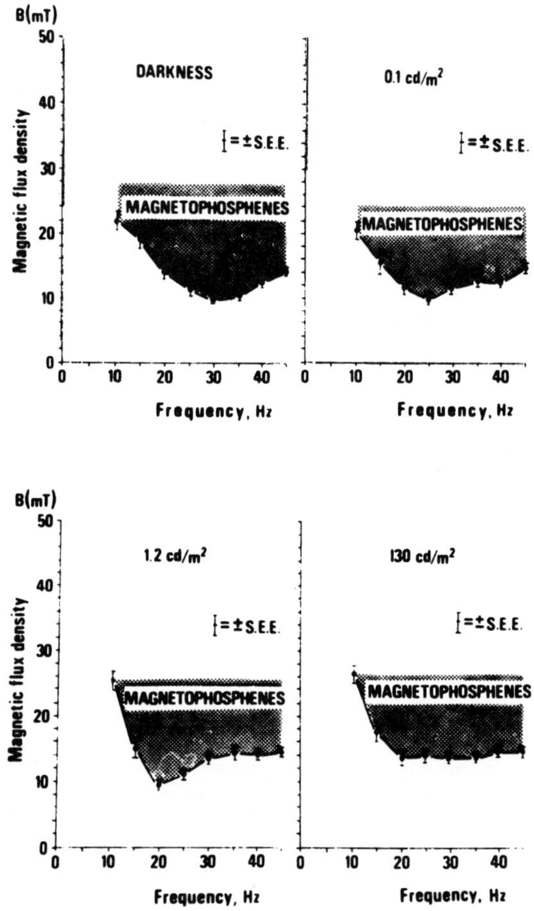

Figure 8. Threshold values for group 1 (11 volunteers) with broad-spectrum light at luminance-level darkness and 0.1, 1.2 and 130 cd/m^2.

Figure 8. One set of results from 10 normal individuals is presented in Figure 8, starting during complete darkness and with increasing background light intensity of broad spectrum light. The frequency of the magnetic field is plotted on the X-axis and the magnetic flux density required to elicit phosphenes is plotted on the Y-axis. A typical response is that the threshold decreases with the frequency of around 20 - 30 Hz. At slightly higher frequencies we often see an increased threshold level. With this response we could see that, at minimum, we would need about 10 millitesla at 20 to 30 Hz to elicit magnetophosphenes.

We also studied persons with normal colour-vision and compared them to people with red-green colour blindness. We found a considerable difference in the threshold patterns. We suggested that these differences discovered via magnetic stimulation could be used to diagnose the red-green blindness. Probably the threshold patterns reflect defects of the retina.

Figure 9. In the upper part of Figure 9, the normal responses to magnetic stimulation are displayed, and on the right side of the figure, the results from the red-green blind persons are shown. Three different wavelengths of background light were used (572, 531 and 443 nm). As can be seen, the threshold curve varies considerably with background light wavelengths and between normal and colour-blind persons.

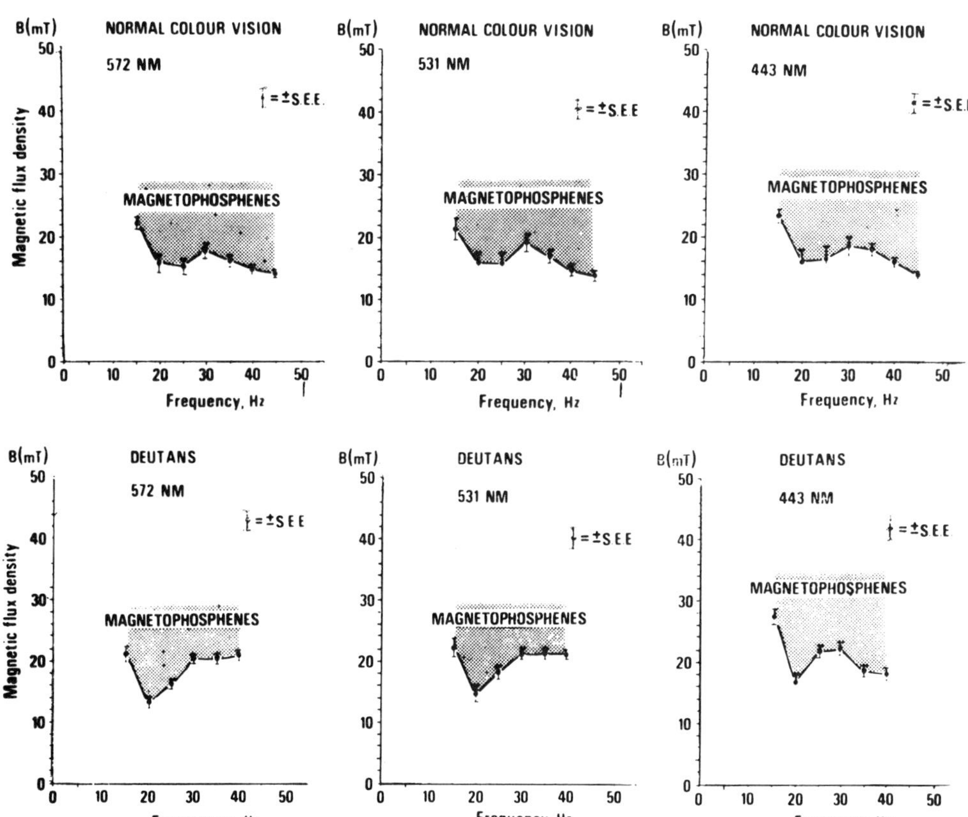

Figure 9. Upper threshold values for group III (six volunteers with normal colour vision) at a luminance of 3 cd/m^2 and wavelengths 443, 531 and 572 nm. Lower threshold values for group IV (nine colour-defective volunteers, deutans) at a luminance of 3 cd/m^2 and wavelengths 443, 531 and 572 nm.

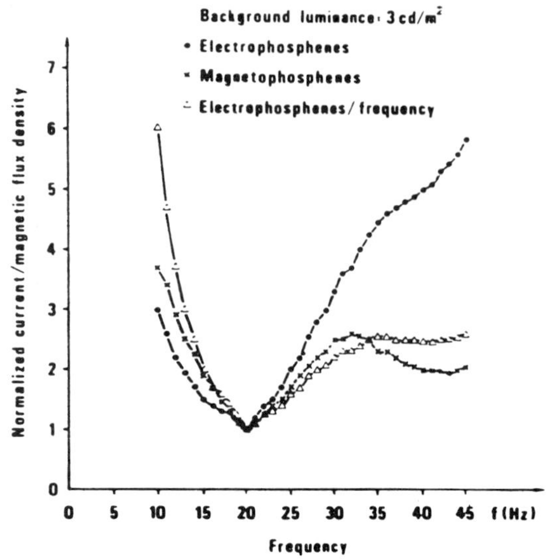

Figure 10. Threshold value curves for magnetophosphenes and electrophosphenes with broad spectrum lighting at a luminance of 3 cd/m^2 and an additional curve where the threshold values for electrophosphenes have been divided by the frequency.

We compared the threshold curves obtained with magnetic stimulation to those obtained when an electric current is passing between electrodes positioned on the temples.

We used the set-up described in Figure 7. The subject under study was galvanically isolated from the generator for safety reasons. We could elicit the same type of response using an electrical current in the same frequency range.

Figure 10. The type of data we obtained is shown in Figure 10. Threshold data for electrophosphenes are compared with magnetophosphene data. After having compensated the electrical current response according to the induction law, we found that the threshold curves for the two stimulation forms are very similar, indicating that the same stimulation mechanism may be behind the two responses.

The typical sensitivity maximum at 20 Hz for both stimulation types can be demonstrated, as you can see in our data. The explanation of the shape of the curve may be the following: the low-frequency part of the curve (high threshold currents) has its shape because the frequency is very low and the stimulation current is passing mainly in the extracellular space of the retina. The high frequency part probably has to do with increasing summation time in neurones at higher frequencies.

FROG RETINA STUDIES

We later became interested in explaining the generation site and mechanism for magnetophosphenes, so we designed a study of the frog retina under low-frequency, magnetic field exposure.

Figure 11. The set-up we used is shown in Figure 11. The frog retina is placed in a holder which is perfused with Ringer's solution. We stimulated the retina with low-frequency magnetic fields of 0 to 40 mT. We could also stimulate the retina with an optical system and we could record the electrical activity from the ganglion-cell layer of the retina with a microelectrode technique.

The protocol for our studies was the following: by stimulating with light and switching the light off and on and adjusting the position of the electrode, we found the ganglion cells that responded to light (so-called on-and-off cells). In addition, we tested the magnetic field response of the same ganglion cells. Fairly many of the ganglion cells sensitive to light did not respond to the magnetic field stimulation. Some did, however, and we could use the retina as a model for explaining the magnetophosphene phenomenon.

Figure 12. This is a more complete block diagram of the set-up showing the various technology used: the filtering techniques, the impedance probe, the timer for the magnetic field exposure of the retina, the sine wave generator and the frequency counter are shown. Ganglion cell responses could be recorded on a tape recorder for later computer analysis.

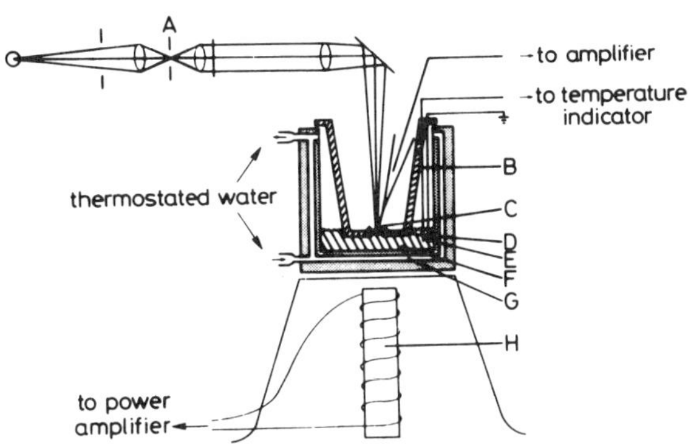

Figure 11. The experimental arrangement. A: optical system, B: thermocouple, C: retina, D: preparation support, E: indifferent electrode, F: Ringer's solution, G: thermo-regulated water +10° C, H: electromagnet.

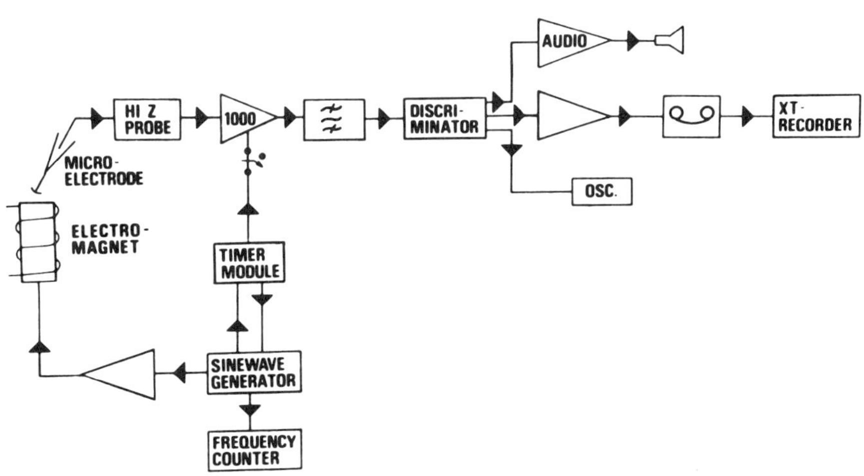

Figure 12. Microelectrode amplifier, recording system and electromagnetic excitation arrangement.

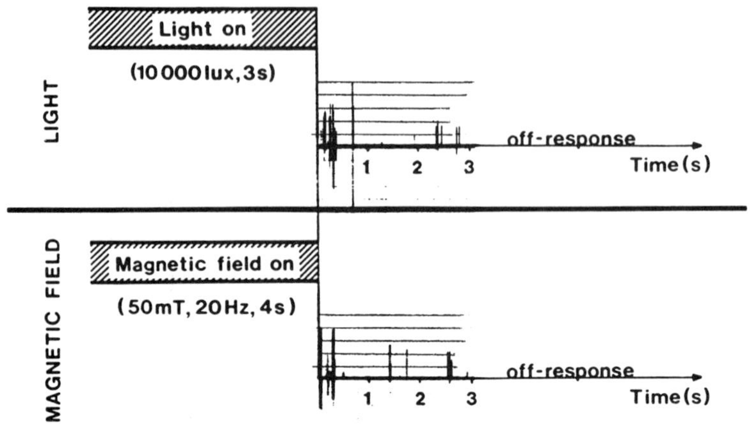

Figure 13. The electrical activity from an on/off-cell after cessation of stimulation with light and magnetic field, respectively.

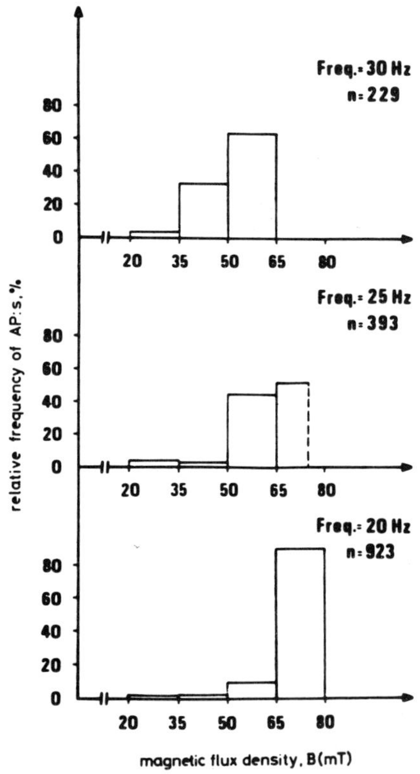

Figure 14. Relative frequency of action potentials (AP:s) from the ganglion cells the first second after switching off a magnetic field at different flux densities and with the frequencies 20, 25 and 30 Hz respectively. (Normalised to the sum of all AP:s at each of the three frequencies.)

Figure 13. This figure shows the results we obtained after having stimulated certain ganglion cells partly with light but also with a magnetic field. The upper part of the figure shows the off-response of ganglion cells after 3 seconds of light stimulation. The same cells were then exposed to the 20 Hz magnetic field for 4 seconds. The off-response is similar to the response obtained with light stimulation. The intensity and the length of the action potential train increased in length when we increased the flux density of the field. This "after-response" sometimes lasted up to a whole minute after the magnetic field was switched off, which may correspond to the "after-images" we experienced in the magnetophosphene studies. The mechanism behind the stimulatory effect of the magnetic field is probably the generation of eddy currents in the complex nervous system of the retina.

Figure 14. This figure shows the number of action potentials of the ganglion cell response following the increase of flux density of the magnetic field. Three different field frequencies have been used (20, 25 and 30 Hz). There seems to be a kind of threshold around 30-50 millitesla above which the number of action potentials generated increases dramatically. Again, 20 Hz was the most effective stimulation frequency.

HUMAN SKIN STUDIES

The last of the three projects I will present today regards stimulation of human skin with fairly strong magnetic fields in the kilohertz range. The studies were performed together with Professor Ueno. We were asked by a company in Sweden that used strong audio frequency magnetic fields for industrial purposes to study whether such fields could stimulate the skin or some parts of the neural system of industrial workers. The workers indicated that they were "sensing" the field when operating the machines generating the fields. In order to study the problem, we started with experiments as shown in Figure 15.

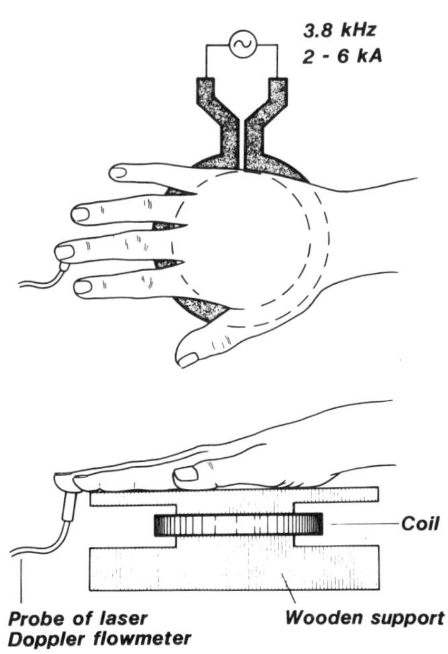

Figure 15. Magnetic stimulation of the hand. The hand was fixed on a wooden support which was inserted in the induction coil without mechanical contact. The hand was exposed to alternating magnetic fields, energising the induction coil by means of alternating currents of 2-6 kA at 3.8 kHz. Blood flow of the middle fingertip was measured by a laser Doppler flowmeter before, during and after magnetic field exposure.

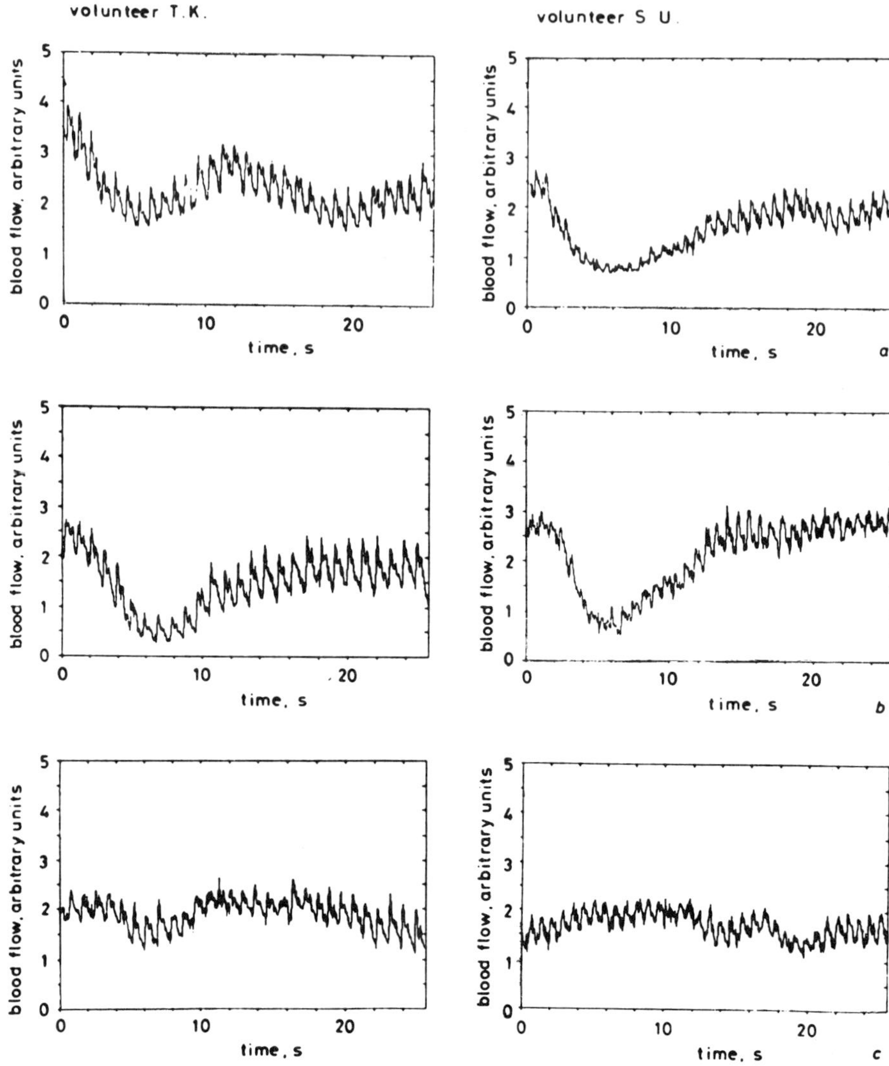

Figure 16. Blood flow changes during magnetic field exposures. The left column shows the results from one volunteer, and the right column shows the results from another volunteer. The magnetic fields were applied at t = 0. During the whole period the hand was continuously exposed to the magnetic fields. The applied currents to the coil and the corresponding magnetic flux densities at the hand were (a) 6 kA, 48 mT, (b) 4 kA, 32 mT and (c) 2 kA, 16 mT. The pulsatile character of the skin blood flow can be seen in the recordings. Even the dicrotic notch can be clearly seen.

Figure 15. A coil monitored close to the hand was fed with AC currents 2-6 kA at 3.8 kHz. A laser Doppler blood flow meter was attached to one of the finger tips. This method estimates blood flow in the microcirculation of the skin.

Figure 16. In Figure 16, responses in skin blood flow following magnetic exposure are shown for two subjects at three magnetic flux densities (48, 32 and 16 mT). The figure shows the blood flow changes when the stimulation and the blood flow measurements take place on the same hand. A rapid and large decrease in flow is followed by a period of slightly suppressed flow levels at the highest magnetic flux density levels. At 2 kiloamperes we see a very small response, if any. The explanation for the blood flow changes following magnetic field exposure is that nervous information from the skin sensory receptors is sent via the spinal cord to the central nervous system, which modulates the efferent nerve activity and affects the blood flow of the skin. This reflex loop is schematically shown in Figure 17.

Figure 17. These were the three magnetic stimulation projects I wanted to discuss with you this afternoon. I will be happy to respond to questions. Thank you very much.

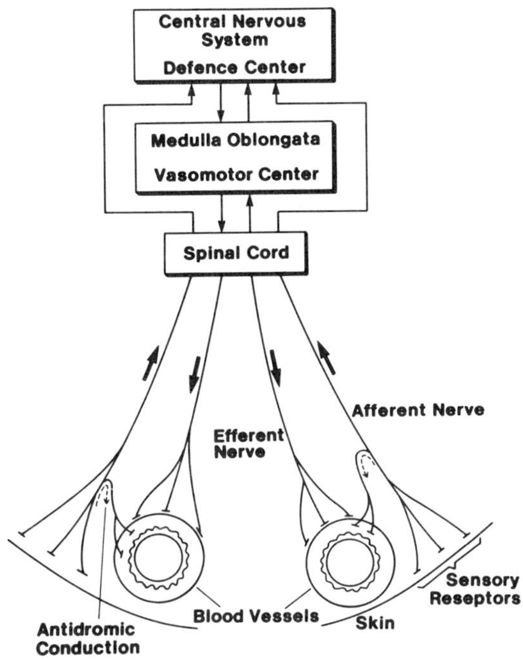

Figure 17. Schematic diagram of innervation of the peripheral vascular system, showing the mechanism of the effect of alternating magnetic fields on the skin blood flow.

DISCUSSION

Chairman Prof. I.L. Freeston: Thank you very much, Professor Öberg. I might add that in the work that we described this morning on magnetic stimulation, we were in fact aware of the work that Professor Öberg's group had been doing. But, as he described, it was invasive and we didn't think we had the elegant experimental technique that he obviously had and we concentrated very much on non-invasive techniques. And so we perhaps took a different route than he did. I think we've got time for perhaps one or two very short questions.

Prof. M. Markov: I was very happy to see some slides demonstrating what I call a "biological window" with respect to frequency and amplitude. My question is, a little bit of an extension probably from what I saw, there was a report from early astronauts that they saw magnetophosphene reactions. Can you comment on this?

Prof. Öberg: No, it could not have been magnetophosphenes that they saw because fairly high flux densities are required to elicit magnetophosphenes. However, phosphenes can be generated via other mechanisms; for instance, pressure differences. Pressure against the eyelobe gives rise to phosphenes similar to the ones generated by the magnetic field.

Prof. Freeston: Any other comments or questions? Then, thank you very much, Professor Öberg.

MAGNETIC STIMULATION OF PERIPHERAL AXONS: MODELS AND EXPERIMENTS

Peter J. Basser

National Institutes of Health
Bldg. 13, Rm. 3W13
Bethesda, MD 20892
USA

INTRODUCTION

Although clinical magnetic stimulation was demonstrated a decade ago [1, 2, 3], little work has been done since then to elucidate the underlying physical principles that govern nerve excitation by electromagnetic fields [4]. The potential benefits of this understanding are numerous. A mathematical model of magnetic stimulation could be used to design improved hardware (e.g., coils and current sources) and to provide the expressions with which to perform this optimization. A model would also permit us to design experiments to test its own predictions, to reconcile observations in numerous descriptive experiments, and to suggest unanticipated sources of experimental error. It could also be used to predict the site of stimulation along an axon and to derive clinically useful scaling laws.

Since the diagnostic and therapeutic benefits of magnetic stimulation have already been addressed, I will focus on the physical underpinnings of magnetic stimulation; specifically, I will show how the equations of electromagnetic induction and an electrical cable equation can be combined to explain excitation of long myelinated axons.

Figure 1 is a schematic diagram showing a magnetic stimulator above a peripheral axon in a limb. The stimulator is represented by a circular inductive coil in series with an RC circuit that represents the current source. When the switch (marked "S") is thrown to the left, the capacitor charges to a voltage V_0. When the switch is thrown to the right, current flows to the ground through the inductive stimulating coil and the resistor. Under "appropriate conditions," an axon will be excited. But what are these conditions?

THE MODEL

To answer this question, we employ basic physical principles, particularly Maxwell's equations. The electric field arises from two sources, charge distributions and time-varying currents:

$$\mathbf{E} = -\nabla\phi - \frac{\partial \mathbf{A}}{\partial t}.$$

Above, $\mathbf{E}$ is the net applied electric field, ϕ is the electrostatic potential, and $\mathbf{A}$ is the magnetic vector potential. In magnetic stimulation, electric fields are induced by the time-varying

Biomagnetic Stimulation, Edited by
S. Ueno, Plenum Press, New York, 1994

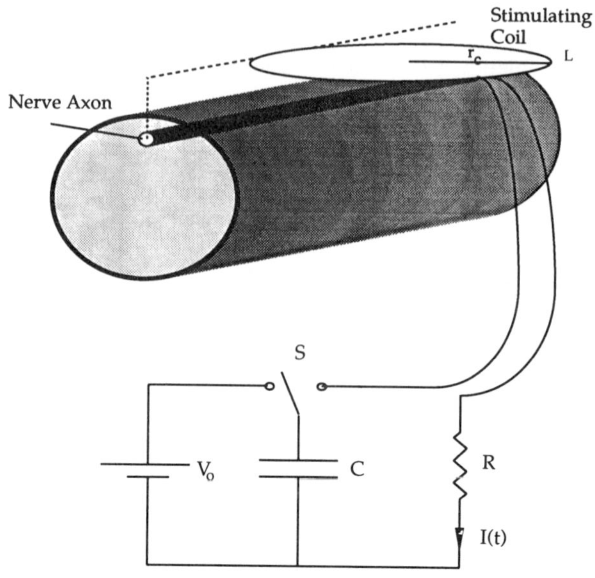

Figure 1. A schematic diagram of the cylindrical limb, nerve axon, and stimulating circuit. The coil radius, r_c, is 4.5 cm and the limb radius is 3 cm. When S is switched to the left, the capacitor C charges to a voltage V_o. When S is to the right, the capacitor discharges through the coil of resistance R and inductance L, creating the current pulse I(t) in the coil.

current in the coil and by charge accumulation at tissue boundaries. The net applied field in the hypothetical limb can be determined by a three-step process. First, we evaluate $\partial \mathbf{A}/\partial t$ from the coil's geometry and current density. Second, we determine the induced charge density that develops at the tissue/air interface to cancel the normal component of the induced electric field there. Third, we determine the electrostatic potential (and its gradient) in the limb that results from this interfacial induced charge distribution by solving Laplace's equation [5].

Having determined **E** in the tissue, we still have to show how it acts to stimulate a nerve. To elucidate this relationship, we rederive the cable equation for the myelinated axon, taking into account the additional net applied electric field impressed upon the nerve [6, 7, 8]. This method is slightly more cumbersome for myelinated axons than for unmyelinated axons, since we must consider separately the effects of the net applied electric field on the active nodal regions and on the passive internodal regions (shown schematically by the horizontal stripes in Figure 2) [6,7].

The myelinated axon is modeled as a distributed, composite medium [9, 10], as shown in Figure 3. The description of the nodes of Ranvier [11, 12] originates in the work of Hodgkin and Huxley. Here, each node is modeled as a discrete membrane current source contributing to the total membrane current [6]. In the internodal region, the transmembrane potential is governed by a passive cable equation [6]:

$$\lambda_m^2 \frac{\partial^2 V}{\partial x^2} - \tau_m \frac{\partial V}{\partial t} - (V - V_r) = \lambda_m^2 \frac{\partial E_x(x,t)}{\partial x} .$$

This passive cable equation is a parabolic partial differential equation. Its parameters τ_m and λ_m are the myelinated membrane time and space constants, respectively; they represent the temporal and spatial decay to a point perturbation in the transmembrane potential.

A novel feature of this composite cable equation is that it is forced by a source term,

$-\partial E_x(x,t)/\partial x$ [6, 7, 8],

which is the activating function for magnetic stimulation of an infinite axon. It is proportional to the transmembrane current impressed by the net applied electric field, and is analogous to activating functions for stimulation by surface electrodes and current microelectrodes [6].

Finally, in constructing the model of the myelinated axon, we must also incorporate scaling relationships that were first elucidated in the 1930s by Hursh and later formalized by Rushton, Tasaki, FitzHugh, and Goldman [13, 14, 15, 16, 17]. These scaling laws are constraints placed on the ratio of the inner and outer diameters of the myelinated axon, d_i and d_o, and the ratio of the distance between nodes of Ranvier, L, and d_o:

$$\frac{\Lambda}{d_o} = 100 \; ; \; \frac{d_i}{d_o} = 0.6 \; .$$

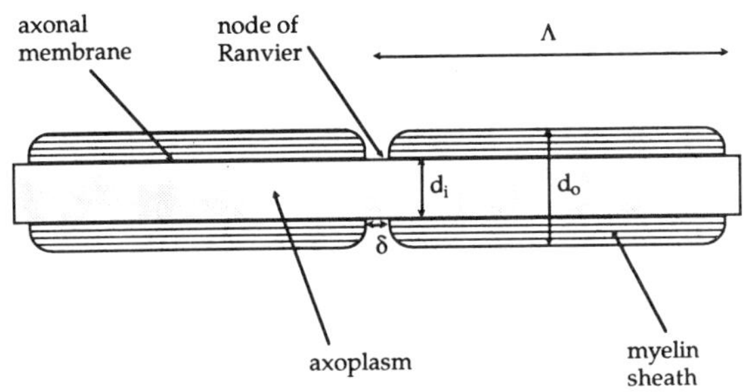

Figure 2. A section of a myelinated nerve axon. The axonal membrane contains active regions, called nodes of Ranvier, that are joined by passive segments insulated by myelin. Nodes are spaced a distance Λ apart and are δ wide. The axon has an outer diameter of d_o and an inner diameter of d_i.

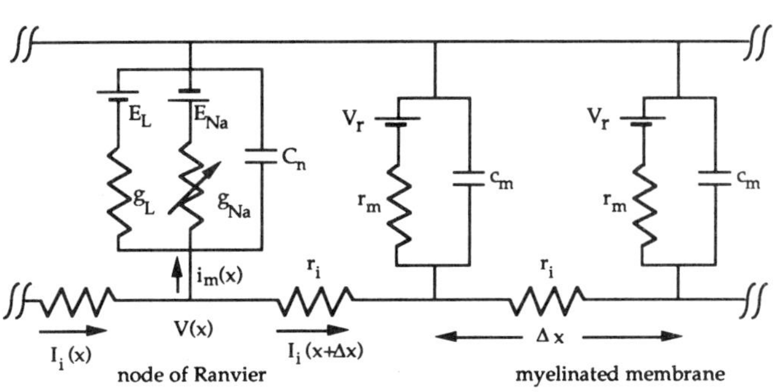

Figure 3. A distributed model of a myelinated axon. At the node the membrane current per unit length, $i_m(x)$, flows through either the nodal membrane capacitance, C_n, or the sodium or leak channels. Ohm's law relates the axial intracellular current, $I_i(x)$, to the intracellular electric field; the equation of continuity relates $I_i(x)$ to the membrane current per unit length, $i_m(x)$. The intracellular resistance per unit length is r_i; the extracellular resistance is zero. The capacitance per unit length of the myelin is c_m; the resistance-length of the myelin is r_m. E_L and E_{Na} are the leakage and sodium Nernst potentials, respectively; g_L and g_{Na} are the leakage and sodium conductances, respectively. The batteries, V_r, represent the action of active pumps and channels beneath the myelin that maintain the resting potential of the membrane at -80 mV.

FINDINGS

The applied electric field in tissue (in the quasi-static limit), and thus the activating function, can be expressed as the product of a function of time, dI/dt, and a function of space that depends only on the geometry of the coil and of the tissue [6]. In Figure 4 we see that for a "monophasic" pulse, dI/dt attains its maximum at $t = 0$, declines rapidly until it crosses zero at $t = t_c$ (about 0.2 msec), overshoots it, and asymptotically approaches zero for large t. Even for simple coil and tissue geometries, the spatial part of $-\partial E_x(x,t)/\partial x$ cannot generally be expressed in closed form (although it can be expressed in terms of elliptic integrals for a circular coil); still, it can be evaluated numerically with little difficulty. For a circular coil, the spatial part of $-\partial E_x(x,t)/\partial x$ varies along the axon. It is an odd function, the maxima and minima of which correspond to anodic and cathodic regions, respectively [8].

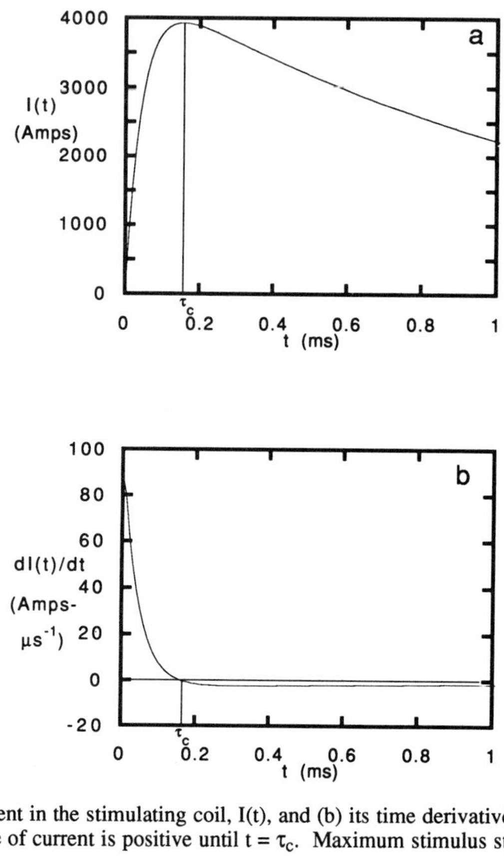

Figure 4. (a) The current in the stimulating coil, I(t), and (b) its time derivative, dI(t)/dt, as functions of time. The rate of change of current is positive until $t = \tau_c$. Maximum stimulus strength is proportional to dI(0)/dt.

Once the position of the axon, its diameter, the coil geometry, and the stimulating circuit characteristics have all been prescribed, it is possible to compute the transmembrane potential as a function of the distance along the axon and of time. The circular coil is assumed to lie just above the nerve, tangent to it, with the center of the coil lying in the plane $x = 0$ (as shown in Figure 1). Figure 5 shows a contour plot of the transmembrane potential, V versus x, the distance along the axon, and t, the time after stimulation. Approximately 0.1 ms after the current starts flowing in the coil, a portion of the nerve to the left of the center of the coil has already become depolarized by about 80 mV above the resting potential. Another 0.3 ms later, leftward- and rightward-running action potentials originating at the center of this region of depolarization are clearly discernible. Meanwhile, to the right of the center of the coil, a portion of the nerve is hyperpolarized by about 5 mV below resting potential. The model predicts this asymmetric behavior, even though the nerve and coil are symmetric about the plane x = 0. The depolarized region behaves as though it were being stimulated by a cathode, while the hyperpolarized region behaves as though it were being stimulated by an anode.

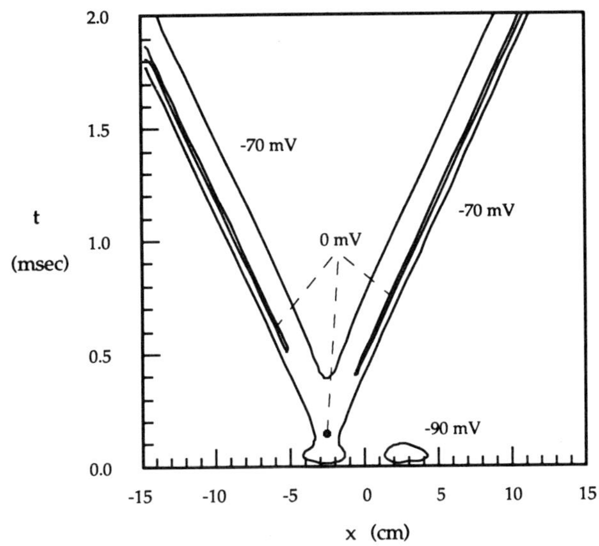

Figure 5. A contour plot of the transmembrane potential along a 20-μm axon located 0.15 cm below the surface of the limb, in response to a suprathreshold stimulus (V_0 = 1600 V). The origin of stimulation is at x = -2.5 cm, while the latency is approximately 0.15 ms.

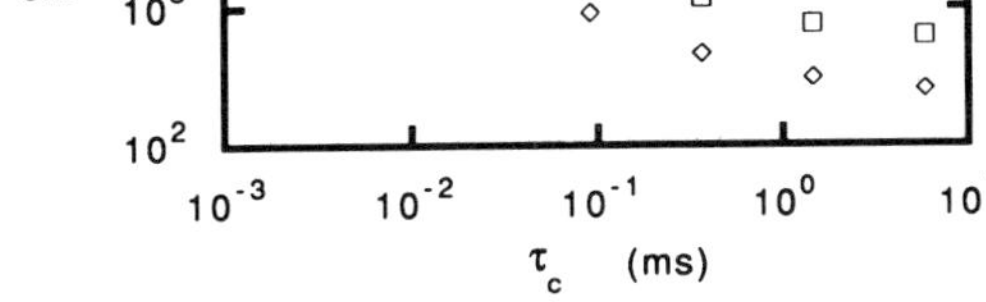

Figure 6. The threshold stimulus strength vs. stimulus duration. This simulation is performed by varying the resistance and capacitance in the stimulating circuit while keeping the damping factor constant. The threshold stimulus strength approaches an asymptote for long-duration pulses, and is inversely proportional to τ_c for short-duration pulses. Smaller-diameter axons require larger threshold stimuli.

The numerically derived strength-duration curve shown in Figure 6 is also revealing. For short pulses, the stimulus strength is approximately inversely proportional to the pulse duration; for long-duration pulses, threshold stimulus strength approaches an asymptotic rheobase level. These model results are qualitatively similar to strength-duration curves for electrical stimulation, except that the breakpoint between long- and short-duration pulses in magnetic stimulation is not as well defined as it is in electrical stimulation, where rectangular current pulses are used [18].

Figure 6 also shows that nerves with outer diameters of 5, 12, and 20 mm have different strength-duration characteristics. Specifically, for the same duration, large (in diameter) nerves are much more easily stimulated than small ones. Figure 7 shows the threshold stimulus strength (now measured in terms of capacitor voltage) for different axon diameters. The numerically generated data are almost perfectly fit to a power law, in which the threshold stimulus strength is inversely proportional to the square of the fiber diameter [6]. In addition, this scaling law can be derived from first principles for an "infinite" axon [6]. We have not yet tested this inverse-square law experimentally, either *in vivo* or *in vitro*.

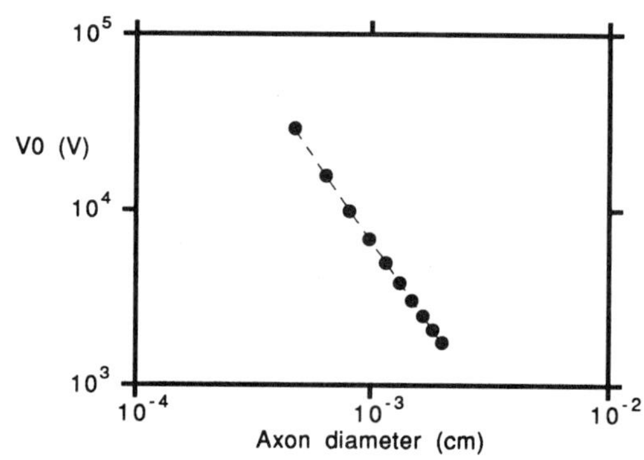

Figure 7. The threshold capacitor voltage required to stimulate axons of different diameters. The stimuli have different amplitudes but the same durations. A regression line was fit to the data; it has a slope of -2.01 and a coefficient of correlation of 0.9997, indicating that threshold voltage is inversely proportional to the square of the axon diameter (R = 0.47 Ω, L = 20 μH, C = 3100 μF, axon depth = 0.15 cm).

Figure 8 shows a computer-generated diagram of the stimulating coil lying above a cylindrical limb, as in Figure 1. A nerve running axially through this uniform volume conductor will be stimulated if it passes through a volume of stimulation -- a region in which the axial electric field gradient is large enough to stimulate an axon of that diameter. The hollowed-out regions in the limb show these depolarized regions for axons 20 mm in diameter or larger [6].

Mathematical models have little value unless they make experimentally verifiable predictions. This model makes several predictions, some of which have already been validated. For example, we have shown that $-\partial E_x(x,t)/\partial x$ is responsible for electromagnetic stimulation of a long axon [19]. The first experiment was performed to determine whether the predicted and observed distances between the electromagnetic cathode and anode agreed with one another. Since the activating function for magnetic stimulation is an odd function, reversing the direction (sign) of the current in the coil exchanges the positions of the electromagnetic anode and cathode, thus causing an accompanying change in the arrival time of an action potential at a distant electrode [8]. (Changing the direction of current in the coil is something we might not consider doing unless we know the form of the activating function.) To determine the distance between the electromagnetic anode and cathode, we multiply the measured delay in arrival of the action potential by the speed of propagation, which can be measured independently by using a small skin electrode.

Figure 9 shows the experimental system. Using a Dantec figure-eight-shaped coil, first developed by Ueno et al. [20], we observed a delay (latency shift) of 0.65 ± 0.05 ms for

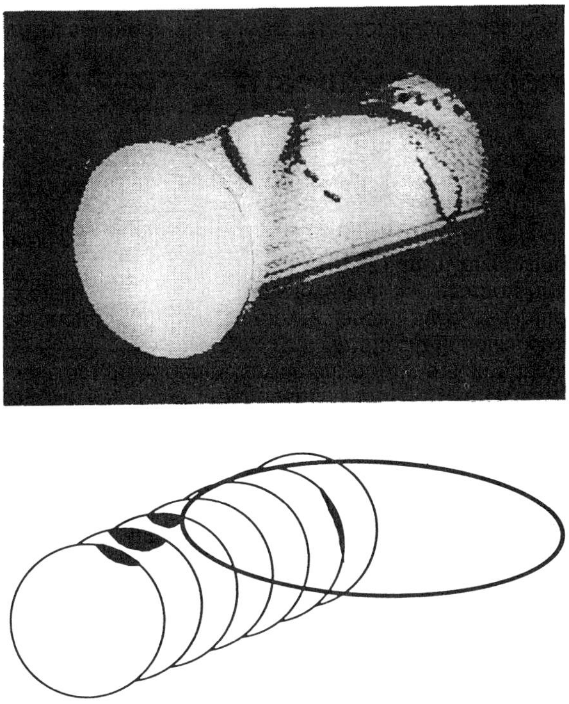

Figure 8. A three-dimensional rendering of the volume of stimulation in a cylindrical limb, produced by a circular coil. The regions in which threshold stimulus strength is greater than 682 mV cm^{-2} are displayed as cavities in this computer generated figure. An axon that is oriented axially, having a diameter greater than or equal to 20 μm, and passing through one of these cavities, is expected to be stimulated.

motor and sensory responses, which corresponded to a space between the anode and the cathode of about 3.7 cm; the predicted separation using the mathematical model is 4.1 ± 0.5 cm [19]. The predicted and experimental results also agree closely with one another in *in vitro* experiments [21]. Differences between theory and experimentation may be ascribed to variations in the site of magnetic stimulation relative to coil position [22].

We have also measured the site of magnetic stimulation with respect to the coil. Once we have found the tract of the nerve using electrical stimulation by measuring the response from a distant electrode on the fingers, we inch a figure-eight coil along the same nerve tract, comparing the measured shape, latency, and amplitude until the electrical and magnetic waveforms superimpose. We can then locate the site of the electromagnetic cathode with respect to the coil. Next, using the model, we calculate the location of the electromagnetic cathode by assuming that it is localized where the activating function is at maximum. For the butterfly coil, the location of the cathode is found to be just distal to the point of contact between the two loops in the coil (as in Figure 10), which is just where we determined it to be experimentally in this inching experiment [19]. This result suggests that $-\partial E_x(x,t)/\partial x$ is the quantity controlling the stimulation, at least for long peripheral nerves. We make no claims about cortical nerves, however, as they have a different architecture.

Admittedly, the numerical model that was used to predict the nerve's behavior is still quite complicated. It is based on Maxwell's equations on microscopic scale, involving nerve membrane properties and anatomy that we can only estimate from the literature. Generally, many of these quantities we cannot measure *in vivo*. Furthermore, the number of simultaneous differential equations we must solve adds additional complexity and time, requiring a supercomputer to determine the gross behavior. It is therefore prudent to reduce the number of parameters and equations while maintaining the generality and utility of the model. One

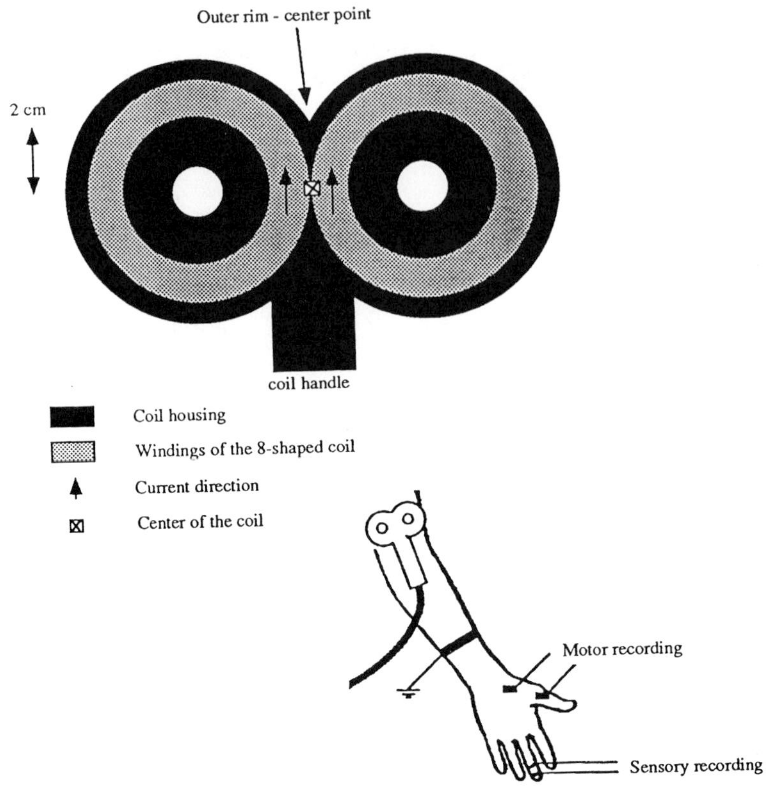

Figure 9. Configuration of Dantec figure-eight coil, limb, and measurement electrodes.

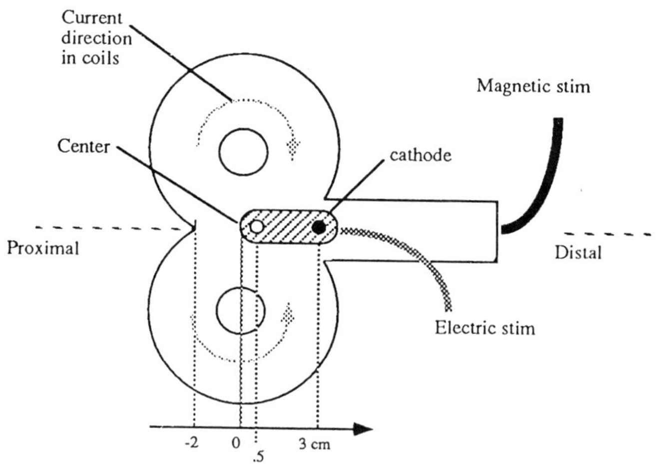

Figure 10. The electromagnetic and electrical cathodes superimposed.

clue to simplifying the model is that the nerve acts electrotonically until threshold is reached. Thus, it is wise to study the response of a passive myelinated axon to electromagnetic stimulation [9, 23, 24].

We have derived a cable equation that is appropriate for subthreshold stimulation of a myelinated axon using homogenization methods. This cable equation has only two unknown parameters, the effective space and time constants, λ_e and τ_e, respectively, which depend on the microscopic structure of the axon, the resistivities, and the capacitances. This two-parameter nerve model obviates knowledge of the precise value of other geometric quantities, such as the thickness of the node of Ranvier, the separation of nodes, and the axoplasmic resistivity, as well as the kinetic parameters -- quantities that cannot be determined *in vivo*, noninvasively, or cost-effectively at this time.

This model is further simplified when we realize that the characteristic length over which the electric field varies using existing stimulating coils is more than an order of magnitude larger than the effective space constant of the myelinated axon, so that the myelinated axon is effectively space-clamped. This permits us to derive analytic solutions for the transmembrane potential as a function of time in the stimulated region, provided that subthreshold conditions are maintained. Analytical and numerical model results are compared in Figure 11. The numerically calculated points for 5-, 12.5-, and 20-mm fibers lie close to the curve predicted by the analytical model. This subthreshold model of the axon can be used to design more focused stimulators [23].

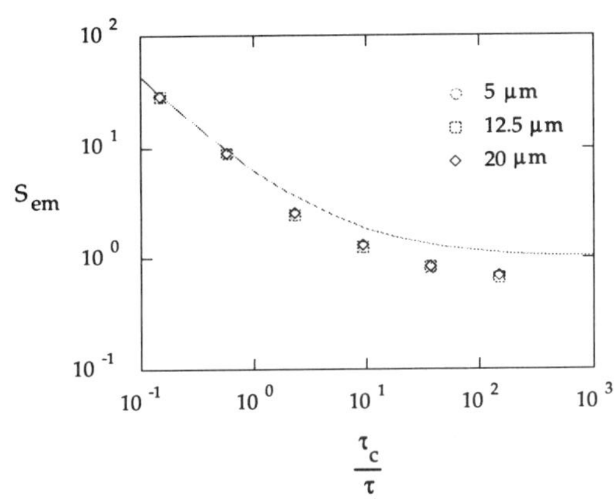

Figure 11. The dimensionless electromagnetic stimulation number, S_{em}, at threshold vs. the normalized pulse period, τ_c/τ. The data points are obtained from numerical computations using the full model. Superimposed is the solution to the homogenized, passive cable model. This curve summarizes the results of many numerical simulations in which axon diameter and pulse duration are varied, and establishes the utility of the homogenized model for describing near-threshold events in electromagnetic stimulation.

CONCLUSION

We have shown that low-frequency electromagnetic fields stimulate long axons in the same way as electrostatic fields -- by causing current to flow across the axon's membrane. By exploiting the analogy between nerve stimulation by electrostatic and electromagnetic fields, we can predict how subthreshold electromagnetic stimuli produce an electrotonic response of the axon, how supratheshold stimuli elicit action potentials, and how the location and dimensions of cathodic and anodic regions can be determined. Furthermore, we can predict the strength-versus-duration curves and derive clinically useful scaling relationships [25].

ACKNOWLEDGMENTS

Thanks to Shoogo Ueno for organizing the International Symposium on Biomagnetic Stimulation, at which this work was presented, and to Brad Roth and Barry Bowman for editing this manuscript.

This chapter was prepared at Bethesda, Maryland with the sponsorship of the National Institutes of Health. Reproduction for the purposes of the United States Government is permitted.

REFERENCES

1. A.T. Barker, R. Jalinous and I.L. Freeston, Non-invasive magnetic stimulation of human motor cortex, *Lancet* ii: 1106-1107 (1985).
2. A.T. Barker, I.L. Freeston, R. Jalinous and J.A. Jarratt, Clinical evaluation of conduction time measurement in central motor pathways using magnetic stimulation of human brain, *Lancet* i: 1325-1326 (1986).
3. M.J.R. Polson, A.T. Barker and I.L. Freeston, Stimulation of nerve trunks with time-varying magnetic fields, *Med. & Biol. Eng. & Comp.* 20: 243-244 (1982).
4. K.P. Esselle and M.A. Stuchly, Neural stimulation with magnetic fields: analysis of induced electric fields, *IEEE Trans. Biomed. Eng.* 39: 693-700 (1992).
5. B.J. Roth, L.G. Cohen, M. Hallett, W. Friauf and P.J. Basser, A theoretical calculation of the electric field induced by magnetic stimulation of a peripheral nerve, *Muscle & Nerve* 13: 734-741 (1990).
6. P.J. Basser and B.J. Roth, Stimulation of a myelinated nerve axon by electromagnetic induction, *Med. & Biol. Eng. & Comput.* 29: 261-268 (1991).
7. P.J. Basser, R.S. Wijesinghe and B.J. Roth, The activating function for magnetic stimulation derived from a three-dimensional volume conductor model, *IEEE Trans. Biomed. Eng.* 39: 1207-1210 (1992).
8. B.J. Roth and P.J. Basser, A model of the stimulation of a nerve fiber by electromagnetic induction, *IEEE Trans. Biomed. Eng.* 37: 588-597 (1990).
9. P.J. Basser, A cable equation for a myelinated axon derived from its microstructure, (in press).
10. P.J. Basser and B.J. Roth, Electromagnetic stimulation of a myelinated axon, *in:* "Proceedings of the 16th Annual Northeast Bioengineering Conference," Pennsylvania State University, pp.129-130 (1990).
11. S.Y. Chiu, J.M. Ritchie, R.B. Rogart and D. Stagg, A quantitative description of membrane currents in rabbit myelinated nerve, *J. Physiol. (London)* 292: 149-166 (1979).
12. S.Y. Chiu and J.M. Ritchie, On the physiological role of internodal potassium channels and the security of conduction in myelinated nerve fibres, *Proc. R. Soc. Lond. B* 220: 415-422 (1984).
13. L. Goldman and J.S. Albus, Computation of impulse conduction in myelinated fibers: Theoretical basis of the velocity-diameter relation, *Biophys. J.* 8: 596-607 (1968).
14. R. FitzHugh, Dimensional analysis of nerve models, *J. Theor. Biol.* 40: 517-541 (1973).
15. J.B. Hursh, Conduction velocity and diameter of nerve fibers, *Am. J. Physiol.* 127: 131-139 (1939).
16. W.A.H. Rushton, A theory of the effects of fibre size in medullated nerve, *J. Physiol. (London)* 115: 101-122 (1951).
17. I. Tasaki, New measurements of the capacity and the resistance of the myelin sheath and the nodal membrane of the isolated frog nerve fiber, *Am. J. Physiol.* 181: 639-650 (1955).
18. M.J. Panizza, J.B. Nilsson, B.J. Roth, P.J. Basser and M. Hallett, Relevance of stimulus duration for activation of motor and sensory fibers: implications for the study of H-reflexes and magnetic stimulation, *Electroenceph. clin. Neurophysiol.* 85: 22-29 (1992).
19. J. Nilsson, M.B. Panizza, B.J. Roth, P.J. Basser, L.G. Cohen, G. Caruso and M. Hallett, Determining the site of stimulation during magnetic stimulation of a peripheral nerve, *Electroenceph. clin. Neurophysiol.* 85: 253-264 (1992).
20. S. Ueno, T. Tashiro and K. Harada, Localized stimulation of neural tissues in the brain by means of paired configuration of time-varying magnetic fields, *J. Appl. Phys.* 64: 5862-5864 (1988).
21. P.J. Maccabbee, V.E. Amassian, L. Eberle and R.Q. Cracco, Magnetic stimulation of straight and bent amphibian and mammalian peripheral nerve in vitro: Locus of excitation, *J. Physiol. (Lond.)* (1993 -- in press).
22. P.J. Basser, Focal magnetic stimulation of an axon, *IEEE Trans. Biomed. Eng.* (in press).
23. P.J. Basser, An analytical model of magnetic stimulation of a peripheral axon, *in:* "Proceedings of the International Symposium on Biomagnetic Stimulation," Kyushu University, Fukuoka, Japan, pp. 25-27 (1991).
24. P.J. Basser, Model of magnetic stimulation of a myelinated axon, (1991 -- submitted.)
25. P.J. Basser and B.J. Roth, Scaling laws for electromagnetic stimulation of an axon, *in:* Proceedings of the 12th Annual International Conference of the IEEE EMBS, Philadelphia, Pennsylvania, pp. 2238-2239 (1990).

DISCUSSION

Chairman Prof. I.L. Freeston: Thank you very much, Dr. Basser. You really did use your allotted time, your 25 minutes: we started a little late. But, I thought I'd keep the pressure on you.

Dr. Basser: Well, you know I need it.

Prof. Freeston: I'm very pleased to see you started to do theoretical work which will enable people to begin optimizing the process which has, to a large extent, proceeded empirically. Do your calculations take into account the fact that there are bones within limbs and that the geometry is not uniform? Do you have any comments?

Dr. Basser: We're starting to do a 3-D finite element model with a group at Johns Hopkins. There are some models, for instance, of magnetic stimulation in the spine, which indicate focusing near the spinal cord. You could imagine that at the boundary between bone and other tissues, microscopic current density could be changed by a factor of two or three; we think that the effects could be substantial.

On the other hand, we've been restricting our experiments to the tibial nerve or the ulnar nerve where we're far from bone and where these effects don't seem to play an important role. However, ultimately including variations in conductivity will enhance these kinds of modeling efforts.

Dr. E. Toyoshima: Do you have a similar model of the muscle fiber?

Dr. Basser: Well, I think that there are composite models of muscle. In fact, the composite cable equation that I used is very similar to a model that had been developed for muscle, but the membrane properties are substantially different. We haven't really been working with muscle because, as John pointed out previously, it's somewhat more difficult to stimulate. We haven't been able to achieve muscle stimulation with the clinical magnetic stimulators that we've been using at NIH. Perhaps with stronger stimulators we may start to see muscle twitching. I also think that you could probably use a model very similar to the one I presented here except that you have to use different values of λ and τ because capacitances and resistances are different in muscle.

Dr. Y. Mano: Thank you very much. Very nice presentation. I am especially surprised about your determination of size and threshold observations. I have one question. I am a clinician, so when we see a patient with degenerated nerve fiber we have a very difficult time differentiating between the degenerated small nerve fiber and the normal small nerve fibers. Using your technique, is there any difference you can explain?

Dr. Basser: Well, I think that I can say a bit about it. You know that an important application of magnetic stimulation is in the demyelinating diseases, such as multiple sclerosis, which was so eloquently discussed before. One would expect a difference in the wave speed when the resistance of the myelin starts to go down. This may be the best indication that you can obtain from electrical stimulation. I also think that once you know better where the stimulation occurs in magnetic stimulation you can make more accurate measurements in conduction velocities. In the future, you'll hopefully be able to use this method, as well as electrical stimulation, in order to determine the actual velocity of conduction. I would add that conduction velocity is probably still the best indicator of the demyelinating diseases which I think you're referring to.

Prof. Freeston: Thank you very much.

SYMPOSIUM SPEAKERS

Dr. A.T. Barker
Department of Medical Physics
and Clinical Engineering
Royal Hallamshire Hospital
Sheffield
England

Dr. P.J. Basser
Biomedical Engineering
and Instrumentation
Division of Research Services
National Institute of Health
Bethesda, Maryland
United States of America

Prof. I.L. Freeston
Department of Electronic
and Electrical Engineering
University of Sheffield
Mappin Street
Sheffield
England

Dr. K.H. Kraus
School of Veterinary Medicine
Tufts University
North Grafton, Massachusetts
United States of America

Dr. W.J. Levy
Department of Neurological Surgery
Cleveland Clinic Florida
Fort Lauderdale, Florida
United States of America

Prof. M. Markov
Bioelectrochemical Laboratory
Department of Orthopedics
Mount Sinai Medical School
New York, New York
United States of America

Dr. J.A. Nyenhuis
Hillenbrand Biomedical Engineering
Center
Purdue University
Electrical Engineering Building
West Lafayette, Indiana
United States of America

Prof. P.Å. Öberg
Department of Biomedical Engineering
University Hospital
Linköping University
Linköping
Sweden

Dr. J.C. Rothwell
MRC Human Movement and Balance Unit
Institute of Neurology
Queen Square, London
England

Prof. S. Ueno
Biomagnetics and Biosystems Laboratory
Department of Computer Science
and Communication Engineering
Kyushu University
Fukuoka
Japan

INDEX